Quality of Selected Fruits and Vegetables of North America

Quality of Selected Fruits and Vegetables of North America

Roy Teranishi, Editor
U.S. Department of Agriculture

Heriberto Barrera-Benitez, Editor
Comision Nacional de Fruticultura, SARH

Based on a symposium
sponsored by the
Division of Agricultural and
Food Chemistry at the
Second Chemical Congress of the
North American Continent
(180th ACS National Meeting),
Las Vegas, Nevada,
August 24–29, 1980.

ACS SYMPOSIUM SERIES 170

AMERICAN CHEMICAL SOCIETY
WASHINGTON, D. C. 1981

Library of Congress CIP Data

Quality of selected fruits and vegetables of North America.
(ACS symposium series, ISSN 0097-6156; 170)

Mostly papers presented at the Symposium on Quality of Fruits and Vegetables of North America, held during the second Chemical Congress of the North American Continent, which included the 180th National American Chemical Society meeting.

Includes bibliographies and index.

1. Fruit—Quality—Congresses. 2. Vegetables—Quality—Congresses. 3. Fruit—Composition—Congresses. 4. Vegetables—Composition—Congresses. 5. Fruit—North America—Quality—Congresses. 6. Vegetables—North America—Quality—Congresses.
I. Teranishi, Roy, 1922– . II. Barrera-Benitez Heriberto, 1939– . III. American Chemical Society. Division of Agricultural and Food Chemistry. IV. Chemical Congress of the North American Continent (2nd: 1980: Las Vegas, Nev.). V. Symposium on Quality of Fruits and Vegetables of North America (1980: Las Vegas, Nev.). VI. American Chemical Society. VII. Series.

SB360.Q34 635 81-14853
ISBN 0-8412-0662-7 AACR2 ACSMC8 170 1-240
1981

PRINTED IN THE UNITED STATES OF AMERICA

ACS Symposium Series

M. Joan Comstock, *Series Editor*

FOREWORD

The ACS SYMPOSIUM SERIES was founded in 1974 to provide a medium for publishing symposia quickly in book form. The format of the Series parallels that of the continuing ADVANCES IN CHEMISTRY SERIES except that in order to save time the papers are not typeset but are reproduced as they are submitted by the authors in camera-ready form. Papers are reviewed under the supervision of the Editors with the assistance of the Series Advisory Board and are selected to maintain the integrity of the symposia; however, verbatim reproductions of previously published papers are not accepted. Both reviews and reports of research are acceptable since symposia may embrace both types of presentation.

CONTENTS

PREFACE

Although judgments of quality are largely subjective, certain distinctive properties that define the degree of excellence can be identified. Scientists are curious about what specific constituents contribute to the predominant characteristics that determine quality. The ultimate goals of research efforts based on this interest are retention and maximization of foodstuff superiority and suppression of the development of undesirable constituents.

This volume presents discussions of the quality of some fruits and vegetables growing in North America and the processes needed to protect them to bring them to our tables in the best quality possible. Such topics are important because the most nutritious food is absolutely useless if rejected. Furthermore, we must retain the enjoyment of our foods, or life will become dull indeed.

Most chapters are papers presented at the Symposium on Quality of Fruits and Vegetables of North America held during the Second Chemical Congress of the North American Continent, which included the 180th National American Chemical Society meeting. The Congress was jointly sponsored by the American Chemical Society, The Chemical Institute of Canada, and the Sociedad Quimica de Mexico, and was held in Las Vegas, Nevada, August 24–29, 1980. Several chapters were added to make this volume more comprehensive.

ROY TERANISHI
U.S. Department of Agriculture
Western Regional Research Center
Albany, CA 94710

HERIBERTO BARRERA-BENITEZ
Comision Nacional de Fruticultura, SARH
Mexico City, Mexico

March 30, 1981

Quality Factors in California Grapes

A. DINSMOOR WEBB

Department of Viticulture and Enology, University of California, Davis, CA 95616

Definitions of Quality

Everyone knows what quality in a grape or other commodity is--or do they? Quality might be defined as those attributes or characteristics of the grape that make it attractive and pleasant to eat or desirable for conversion to wines which would be pleasant to drink. This definition seems perfectly straightforward and logical until one realizes that those attributes pleasing to one person may be quite unpleasant to others. Economists have defined quality in product or commodity as any characteristics that make it appealing to the purchaser. Here, also, one runs into the difficulty that certain characteristics may motivate one group of people to buy the object, but may be completely unattractive to others. Is there, then, some abstract element of characteristic that can be uniformly recognized as providing relative levels of quality to a substance? Quality in grapes and wines might be defined as that particular balance or harmony of other attributes which makes the grape or wine appealing and interesting to an experienced taster. Realizing that this definition may still be unsatisfactory in that the man in the street may not agree with the expert, we shall nevertheless consider California grapes and the wines made from them from the point of view of this understanding of quality.

Raisin Grapes

The total crop of grapes produced in California varies from year to year according to the vagueries of climate. Further, the percentage of the raisin grapes that are actually dried into raisins varies with the world-wide market for raisins. In years when this market is depressed, sizable quantities of California raisin grape varieties will be used for fresh consumption and for wine production. For example, in 1979, for which preliminary figures are now available, (1) of the total crop of 4,493,000

0097-6156/81/0170-0001$05.00/0

tons, 51% or 2,300,000 tons consisted of raisin grape varieties. Of this quantity of raisin grapes, 60% or 1,374,000 tons was actually dried to produce raisins; the balance being divided among 700,000 tons (30%) crushed, 60,000 tons (3%) canned, and 166,000 tons (7%) consumed as fresh fruit. While very small quantities of Zante currants, Muscat grapes, and some new varieties are dried to make raisins, the very great proportion of the raisin production is from the Thompson Seedless or Sultanina grape variety. There is little question that the best Thompson Seedless raisins are those produced from moderately cropped vines grown on fertile, deep soils (2). Fresh grapes of higher sugar concentration tend to produce raisins that are plumper, sweeter, and more meaty or chewy. Low sugar grapes, in contrast, produce raisins that are less sweet, and which have much larger proportions of skin to flesh. As the sugar concentration in grapes varies inversely with the crop level, economics favors production of raisins somewhat below the maximum potential quality. The commercial vineyard is usually managed to optimize the balance between production and raisin quality.

While there is no question that sugar in the grape is the most important factor in raisin quality, the acid content and the content of other non-volatile solids in the grapes is probably of secondary importance. Acid is particularly important as an aspect of the fresh, fruity flavor. Low acid raisins, just as low acid fruit, taste flat and uninteresting in comparison with examples of sufficient acidity. Flavor compounds in the grape are mainly produced by the dehydrating or drying process through a series of fairly complicated chemical reactions. The contributions of the various products of these reactions to the overall grape flavor is currently under investigation by Singleton and his co-workers (3). It is to be noted that flavors differ significantly between raisins produced by the sun-drying technique as compared with the dehydrator techniques. Further, use of chemicals to break down the bloom layer on the skins and thus facilitate the drying may have a secondary affect on the flavor balance of the particular raisin-type.

Table Grapes

The term "table grape" is used to distinguish fruit that is destined to be eaten fresh rather than to be dried or made into wine. Of course, any ripe grape can be eaten fresh, but over the centuries there has been selection of varieties that were seedless on the one hand and that were larger berried, firmer fleshed and with more attractive colors on the other hand. Selection for attractive color has been paralleled by selection of large size. We find thus that quality attributes in table grapes can be considered from two categories, the visual and the palate. Visual factors of obvious importance to quality in table fruit are size, color, shape, and intact bloom on the berries.

Berry size. There is no question that larger size is a positive factor in favoring sales of table grapes, particularly in the United States market. The validity of this observation is supported by the increase in consumption of Thompson Seedless grapes which have not only been girdled, but also sprayed with gibberellic acid to make the treated berry some four to five times larger than the natural berry (4). For a new table grape variety to be successful in the market, it must be large in size.

Berry color. Table grapes range in color from pale greens through light yellows, golden colors, to the pinks, reds, and nearly black. The mottled or striped light red color on the pale, yellow-green background of the Flame Tokay variety has undoubtedly been one of the factors in making it a success in the market. Similarly, the striking, shiny black and the very large size of the Ribier have undoubtedly been important factors in its sales appeal.

Table grape berry shapes vary much more than do the shapes of the wine or raisin varieties. It is assumed that over the generations interesting mutants of berry shape have been selected and preserved. Table grape berries range from nearly sperical to very elongated, cylindrical forms. Berry shape, as well as color, is certainly one of the factors leading to good sales appeal in these varieties.

Berry surface bloom. Of equal importance to the berry shape and the color is the attractive natural bloom on the surface of the grape berries. The bloom is a thin layer of wax-like material coating the outer surface of the berry and giving the impression that the berry has been lightly and uniformly dusted or frosted (5). Great care is required during harvest and packing of table grape varieties in order that the bloom not be disturbed. Clusters which have been handled roughly show evidence of this in areas where the bloom is removed.

Of particular importance in the table grape varieties that are stored cold for prolonged periods before marketing is the appearance and conditions of the stem. Poor storage conditions can lead to dried out and brownish colored stems which are much less attractive than the turgid and green stems of the fresh fruit. Indeed, stem condition and appearance is one of the more important factors in the overall appeal the cluster of grapes has to the purchaser (6). While visual factors are probably overriding in the customer's decision concerning the purchase of table fruit, palate factors cannot be completely discounted. Here we should consider seedlessness, texture, and flavor.

Seedlessness, particularly for the American market, has become a very desirable attribute in table grapes. In contrast to the practice in Europe, where enormous quantities of the small berried, large seeded Chasselas doré are consumed, preference in the United States is definitely for seedless varieties. Nearly all of the new varieties being introduced are without seeds.

In comparison with the wine grape varieties, table grapes have a very firm and pulpy texture. Most wine grape varieties easily lose their juice when the skin is broken. In contrast, table grape varieties tend to have a texture more like that of an apple or other firm fruit. This characteristic apparently is appreciated by consumers of table grapes who refer to the crispness and firmness of the texture of the berry.

The attribute of flavor is composed of both taste and odor elements. Of the four basic tastes, only those of sweetness and sourness are of much significance. Some table grape varieties, particularly when underripe, may have a hint of bitterness, but the salty taste is almost never apparent. The volatile aroma materials that are perceived in the nasal cavity when the berry is chewed and are thus thought of as part of the taste, are very important to the positive appeal of many table grape varieties. In the Muscat-flavored sorts, with their relative high concentration of the terpene alcohols, (7) flavor is much appreciated. American varieties of the *labrusca* species owe their characteristic aroma to the presence of the ester, methyl anthranilate together with many other organic volatiles. We still do not know what are the important compounds in the characteristic aroma of the Rotundifolia species.

In comparison with the wine varieties, table varieties usually are harvested and consumed at lower concentrations of sugar, typically around 150 grams per liter rather than the 220 to 250 of wine varieties. Acid concentrations, similarly, are usually lower, and in the range of 4 to 6 grams per liter rather than the 8 to 10 of wine varieties. Overall quality in the palate factors of a table grape depends upon a harmonious balance of the elements making up the flavor (8).

Of the total of 4,493,000 tons of grapes produced in California in 1979, 413,000 tons or 9.2% were table grape varieties. Of this 413,000 tons, 225,000 tons or 54.5% were actually consumed as fresh fruit, the balance being crushed to produce wine. It should be noted that it is standard practice world-wide to utilize scarred or blemished table grapes, which cannot be sold for consumption fresh, for the production of standard quality wines or of distilling material.

Wine Grapes

For the 1979 season, of the total of 4,493,000 tons of grapes produced, 1,780,000 tons (40%) were classified as wine grape varieties. Of the wine grape varieties, 1,692,000 tons (95.1%) were crushed to wine or distilling material. It should be noted that the 4.9% not crushed according to the reports actually includes all of the fruit that was consumed fresh and used for all purposes on the vineyard properties.

In great contrast with table grapes, wine grapes do not have their quality influenced by the presence of blemishes or scars produced by weather or rough handling. Quality in wine grapes is determined by the composition of the material inside the berry. Freedom from mold, of course, is a very important factor in wine grape quality just as it is with the other grape types. Wine grapes normally contain 21 to 24% of fermentable sugars--significantly higher than the concentrations usually found in the table grape varieties. Acid levels, also are significantly higher--normally ranging from 6 to 12 grams per liter when expressed as tartaric acid (9). Sugar concentrations within the mentioned ranges are important because the sugar, after conversion to alcohol, determines whether the wine tastes thin and watery or too alcoholic and hot. Similarly, excessive concentrations of acid in the grape lead to wines that are too tart to be palatable, while if the grape is deficient in acid, the wine will be flat and uninteresting.

Aroma compounds. The aroma compounds in the grape are of much importance in determining the wine quality, as well. Recent studies, employing gas chromatographic techniques for separation coupled with mass spectrometric procedures for identification, have permited characterization of some 300 volatile organics as involved in grape and wine aroma and bouquet (10, 11). While a number of these compounds are produced by the yeasts from non-volatiles of the grape, a good many are naturally present in the grape and come through the fermentation either unaltered or with minor modifications. It is interesting that no single compound seems to be completely responsible for the characteristic flavor of any grape variety. However, it is clear that methyl anthranilate is important in the aroma of the *labrusca* species of which the Concord is the most important cultivar. The Muscat varieties of *Vitis vinifera* owe their intense characteristic aromas to the presence of a number of terpene alcohols, among which linalool and geraniol are probably the most important. Cabernet Sauvignon, another *Vitis vinifera* cultivar, has been reported to contain small concentrations of methoxyisobutylpyrazine, a compound with a strong, green bell pepper-like aroma. Even with these three cultivars, many other volatile organics are involved in the complete grape aroma. Most of the other cultivars that have been investigated contain large numbers of volatiles among which no single one has been associated with the characteristic cultivar aroma.

Phenolic compounds. The anthocyanins and tannins are particularly important constituents of wine grapes in that they provide the attractive color to red wines and the desirable bitterness and astringency associated with the table wines. While the concentrations and types of anthocyanins present in the various cultivars and species differ, most red wine grapes contain enough of these compounds in the skins so that satisfactory colors are produced during the fermentation of the wine when the alcohol extracts the pigment from the skins (12). Red vinifera varieties can usually be distinguished from the red varieties of the American species on the basis that the former contain only 3-monoglycoside pigments. As the gene for dyglycoside pigmentation appears to be dominant in most cases, use of the American species in breeding hybrids can usually be detected by this characteristic. Tannins, in contrast with the anthocyanins and pigments, are present in the pulp, seeds, and stems as well as in the skins. Tannin molecules of relatively lower molecular weight are responsible for the bitter taste while those of higher molecular weight give rise to the tactile sensation of astringency or puckeriness. Appropriate levels of these tannins in the wine can be controlled to a certain degree by the conduct of the fermentation, but it is necessary that there be an adequate quantity in the grape in order to have a balanced wine as a result.

Wine grapes must contain a number of growth factors for the yeast in order to support an adequate fermentation to wine. Among these components are a number of vitamins and a number of minerals. In addition to the major requirement for nitrogen, phosphorous, and potassium, yeasts also require a wide range of trace elements. These must all be supplied by the grape in order that the fermentation will be rapid and complete.

Factors Influencing Quality in Grapes

To produce high quality grapes of the three types mentioned above requires careful control of a number of factors among which the variety, the climate, the soil, and vineyard management are probably of greatest significance. There are complex interactions among these factors, and the particular assembly of factors optimum for each of the basic types of grape differs.

Cultivar. It is obvious that the variety or cultivar is of fundamental importance. As mentioned earlier, the composition and size of the grape varieties of the different types vary considerably. This genetic inheritance is fundamental but variations in other factors can influence the expression of the potential quality in the particular cultivar. Thus, while a cultivar such as the Flame Tokay table grape does best in the Lodi region of California, it will produce fruit recognizable as table grapes in many of the other climatic regions of California.

Climate. Of nearly as great importance as the variety or cultivar in determining ultimate quality of the fruit is the climate. Maximum, minimum, and average temperatures and the daily pattern of heat accumulation are very important elements of the overall climate. The different grape cultivars have somewhat different tolerances to temperatures and heat stress, but in general, the grapevine is a temperate region plant--suffering severly from very low winter temperatures and very warm, humid conditions. Rainfall and fog and their distribution through the season are also of considerable importance. Cultivars of the vinifera species need a long, warm, dry growing season. High humidity during the growing season is likely to result in severe infections of fungus diseases which seriously impair the quality of the fruit. Wind can be a problem in that the young tender shoots of the vine in the springtime are easily broken. Of particular danger are late spring frosts which can kill the tender, young shoots of the vine and early winter freezes which can damage the wood of the canes and trunks before they have had time to winter-harden.

Soil. Much importance is attached to the soil of the vineyard by many writers about grapes and wines. Here, as with many things involving living material, moderation seems to yield the most successful results. Excessively fertile soils tend to promote vigorous vine growth with only poor to moderate crops of low quality. On the other extreme, very poor soils which support only very meager leaf growth also lead to uneconomically small crops, although they may be of high potential as far as wine quality is concerned. Most important seems to be that the soil be loose and of a depth such that good drainage prevents the vine's roots staying wet.

Husbandry of the vineyard. Vineyard management, comprising operations such as training, pruning, pest control, and harvesting are obviously of great importance as far as fruit quality is concerned. The varieties of grapevines used in commercial vineyards world-wide have been so highly selected that they require careful training and pruning in order to produce optimum crops of high quality fruit. Left to themselves, they tend to set more fruit than the vine can ripen, and to grow so vigorously that without pruning the vineyard soon becomes an impenetrable bush. The recently conceived ideas of intergrated pest management are proving particularly helpful in controlling the many insects and diseases that attack the grapevine. Although the problems vary with region and from cultivar to cultivar, some effort is required in the management of potential pests in every situation.

The final step in getting high quality fruit from the vineyard to the consumer starts with harvesting. Here, of course, the system and the degree of care varies with the grape type--table fruit requiring much more careful hand labor than does the harvesting of grapes for wine. Of prime importance, however, in all cases is that the harvest be conducted at the optimum time. This means that the concentrations of sugar and acid must be closely monitored toward the end of the ripening period and that the harvest must be conducted in a relatively short period when these constituents are present at their optimum concentrations.

Machine harvesting, while inapplicable to table grape vineyards, is especially suited to raisin and wine grape gathering. In the case of the wine grape, there are the additional advantages that the operation can be conducted during the cool night-time period rather than during the full heat of the day, and that a field crusher can be coordinated with the machine harvestor so that only the must is taken into the winery. With raisin varieties to be sun dried in the vine rows, machine harvesting can be adapted to place the freshly harvested fruit on a continuous paper tray layed in the row following the machine.

Summary

Quality in grapes is somewhat difficult to describe. Something of the economist's definition that quality is any characteristic promoting sale seems present. The idea that quality is an abstract concept based on a balance or harmony of other attributes attractive to the connoisseur also is involved. There are three basic types of grapes: raisin grapes, table grapes, and wine grapes. Quality in raisin grapes seems to be correlated with optimum concentrations of sugar and acid in the fruit at the time of harvest. Visual factors are very important in quality of table grapes. Size, color, shape, and presence of undisturbed bloom are particularly important. Palate factors such as seedlessness and texture as well as a balance of flavor are similarly important. With wine grapes, optimum concentrations of sugar, acid, phenolics, aroma materials, and growth factors are very important.

Factors influencing quality in grapes are variety, climate, soil, and vineyard management.

Abstract

Of the 3,878,000 tons of grapes produced in California in 1978, 64% were converted to wine with the rest divided between raisin and fresh table use. Grape quality factors of importance for raisins are sugar, acid and extract concentrations; for table grapes visual factors such as size, color, shape, and bloom, and palate factors such as texture and flavoring are governing. For wine grapes, appearance is of little importance, but appropriate concentrations of sugar, acid, phenolics, minerals, vitamins, and aroma compounds are critical to high quality in the resultant wine. Grape composition at maturity is governed by the variety planted, the climate (including the temperature regime), the rainfall and fog, the wind, and air pollution. Soil and aspect of vineyard including soil composition, depth, drainage, and vineyard management, including the training system, the level of pruning, pest management, and the timing and method of harvest are also of critical importance. Ultimate wine quality can be no better than potential quality inherent in the particular lot of grapes.

Literature Cited

1. McGregor, R. A., Lockhart, D. "California Grapes, Raisins, and Wine 1979. Production and Marketing" Federal-State Market News Service and California Crop and Livestock Reporting Service: Sacramento, CA, 1980.
2. Winkler, A. J., Cook, J. A., Kliewer, W. M., Lider, L. A. "General Viticulture"; University of California Press: Berkeley, 1974, p 639.
3. Lin, Nen-Hua "Studies on Chemical Brown Pigments and Sensory Characteristics of Raisins Prepared by Different Methods" M.S. Thesis, University of California, Davis, CA, 1980.
4. Weaver, R. J., Sachs, R. M. "Hormonal-Induced Control of Fruit-Set and Berry Enlargement in *Vitis vinifera* L." *In* Biochemistry and Physiology of Plant Growth Substances, Ed. Wightman and Setterfield, Runge Press Ltd.: Ottawa, Canada, 1968, p 957-974.
5. Radler, F. *Am. J. Enol. Vitic.* 1965, *16*, 159-167.
6. Nelson, K. E. "Harvesting and Handling California Table Grapes for Market"; University of California Division of Agricultural Sciences Pub. 4095, 1979.
7. Cordonnier, R. *C. R. Acad. Agric.* 1955, *41*, 399-403.
8. Nelson, K. E. *Am. J. Enol. Vitic.* 1973, *24*, 31-40.
9. Amerine, M. A., Berg, H. W , Kunkee, R. E., Ough, C. S., Singleton, V. L., Webb, A. D. "The Technology of Winemaking, 4th Ed."; Avi Publishing Co. Inc.: Westport, Conn. 1980, Chap. 2.
10. Webb, A. D. "Volatiles in Grapes and Wines"; Proc. Bienn. Int. CODATA Conf. 5th, 1976, Ed. Bertrand Dreyfus, Pergamon, Oxford, England, 1977, p 101-108.
11. Schreier, P. "Flavor Composition of Wines: A Review"; CRC Press: Cleveland, Ohio, 1979.
12. Ribéreau-Gayon, P. "The Chemistry of Red Wine Color" *In* Chemistry of Winemaking, Adv. in Chem. Series 137, A. D. Webb Ed., American Chemical Society, Washington, D.C., 1974, p 50.

RECEIVED April 15, 1981.

2

The Odor Quality of Labrusca Grapes

TERRY E. ACREE

New York State Agricultural Experiment Station, Cornell University, Geneva, NY 14456

Cultivars of the species *Vitis vinifera* are the most prevalent grapes planted in the world. Originally grown in Europe and Asia , vinifera grapes, of which there are several thousand named cultivars(1), also dominate the viticulture of the New World. In North America, however, two other species of *Vitis* (*labrusca*, Bailey and *rotundifolia*, Michaux) are grown in significant quantities(2). The severe climate and virulent disease in central and eastern North America made early cultivation of vinifera grapes largely unsuccessful, at least in the English Colonies. This stimulated the hybridization and cultivation of the more tolerant native grape species for many years.

With the development of methods for disease control and the expansion of viticulture into climates more amenable to vinifera grapes, the percent of native species has decreased to less than five percent of the total North American grape production (3). Although dwarfed by the size of the vinifera grape crop, in excess of 4,000,000 tons, the production of labrusca grapes has increased in the last 30 years. This continued demand for labrusca grapes is due to their superior quality for the production of grape juice and jelly. It is the unique flavor of labrusca grapes, and in particular the cultivar Concord, that is responsible for their superiority. In fact, the flavor of labrusca grapes has become the standard of identity for grape juice and jelly in North America.

THE ODOR OF LABRUSCA GRAPES

The major flavor differences among grape products made with different species of grapes are found in the odor. It is easy to distinguish the odors of grape juices made from vinifera, labrusca, and muscadine species. In contrast, the taste of grapes, that is their sweetness, sourness, and bitterness, is frequently manipulated by processors to the extent that any taste difference among grapes of different species are obliterated.

The one word which has been used for centuries to describe the odor character of native grapes is "foxy". In fact early

0097-6156/81/0170-0011$05.00/0

European visitors to the New World named the native labrusca grapes growing in New England the Northern Fox Grape, and the native grapes growing in the South the Southern Fox Grape. The original meaning of the word "foxy" was explicitly stated in 1722 by Robert Beverly when he described the smell of these grapes as resembling that of a fox(4). This musky animal-like aroma is still considered an undesirable attribute in grapes, but "foxy" must not be confused with the powerful fruity, floral, or candy odors essential to the quality of grape juice and jelly. Even though these odors may not be appreciated in most wine types, they are not the same as foxiness. The cultivars of *labruscana* (Concord, Catawba, Delaware,etc.) were appreciated by early American grape breeders because of their general lack of foxiness. However, Niagara is the one remaining commercial cultivar which has always been considered foxy. Therefore, foxiness should not be considered the dominant odor quality of Labrusca grapes. In this paper, that odor quality common to all labrusca grapes will be referred to as "labrusca character."

During the sensory studies of hundreds of different grape cultivars conducted at the Experiment Station in Geneva, New York, panelists have, with great regularity, identified the presence of labrusca character in the odor of labrusca grapes and often do the same with hybrid crosses between *V. labruscana* and other grape species. Furthermore, they can detect the presence of specific odor components, which in varying degrees, seem to make up the total sensory effect. Four descriptors which are frequently used by these panelists are "foxy","floral", "methyl anthranilate-like", and "cotton candy". It is certainly unwise to assume that these odor components are, in every case, related to single chemical species, but they are probably less complicated in their chemistry than the compound mix that produces the total perception of labrusca character.

Such descriptors are useful in our attempts to sort out the small number of odor-active compounds present in natural products. However, they have little meaning from one laboratory to the next. For example, the aroma described in our work as "cotton candy" appears, for reasons which will be explained later, to be due to the same compound responsible for the strawberry odor detected in the laboratory of Rapp(5) in Germany. The confusion these non-chemical descriptors create, will be minimized once we know the causative agents for odor perception. Then we can use a chemical name to describe that perception. In the meantime, we must tolerate to some extent the use of these words in our day-to-day research.

METHYL ANTHRANILATE

Studies of the volatile composition of grapes have, through the years, revealed the presence of so many odor-active compounds that it is very unlikely that a single compound is responsible for more than a few percent of the total odor character of grapes of any kind(6). Furthermore, among the hundreds of volatiles present in grapes (Figure 1.), only a very few probably have any odor activity. There must exist a number of chemical compounds present in grapes at particular concentrations that when taken together produce the characteristic odor quality of a particular cultivar. Associated with each of these odor-active chemicals is a specific, though not necessarily unique, odor quality.

One such compound is methyl anthranilate, the first compound ever associated with odor character of a particular grape species (7). Originally identified in neroli oil (8), it was used to produce synthetic grape-flavored products before it was observed in grapes. In 1923 Sale and Wilson (9) analyzed the methyl anthranilate content in 55 cultivars of grapes, including both labrusca and vinifera cultivars. They found methyl anthranilate in only 14 of these. In 1976 using gas chromatography, we found methyl anthranilate in only 8 of 45 cultivars (10). Certainly this compound plays an important role in the flavor of some of the labrusca grapes but it is not solely responsible for the labrusca character.

Many wines with a strong labrusca character have been found to have a methyl anthranilate content well below the apparent odor threshold (10). Furthermore, the addition of this compound to vinifera wines at very high concentrations does not produce a labrusca aroma nor does it produce a wine with any foxiness. There are small amounts of other anthranilate esters, ethyl, propyl, *sec*.-butyl and the analogous compound *o*-aminoacetophenone in labrusca grapes but there are no data explaining their precise contribution, if any, to the methyl anthranilate-like aroma.

In the last 60 years, perhaps too much emphasis has been placed on the methyl anthranilate content of labrusca grapes. For example, measurements of this compound have been used to indicate grape maturity(11); it has been used as an indicator of the quality of grape products(12), and as a means of monitoring the performance of essence recovery equipment(13). Furthermore, the amount of methyl anthranilate combined with the volatile ester content of grapes is presently being used to aid the grape breeders in their attempt to produce new cultivars with certain odor characteristics(14).

One of the dangers of placing too much importance on the presence of methyl anthranilate in grape products is that such products may tend to smell more like the simple imitation grape flavors rather than the natural product. It is one of the goals of flavor research to prevent such ironies from occurring unintentionally.

FOXINESS

So far our attempts to isolate a pure chemical component from labrusca or muscadine grapes with a clear foxy odor have been unsucessful. There is, however, one component, *trans*-2-hexen-1-ol, apparently present in all grape species, that has an odor very similar to the foxy odor component of Niagara grapes. Possibly, this compound is present in foxy-smelling

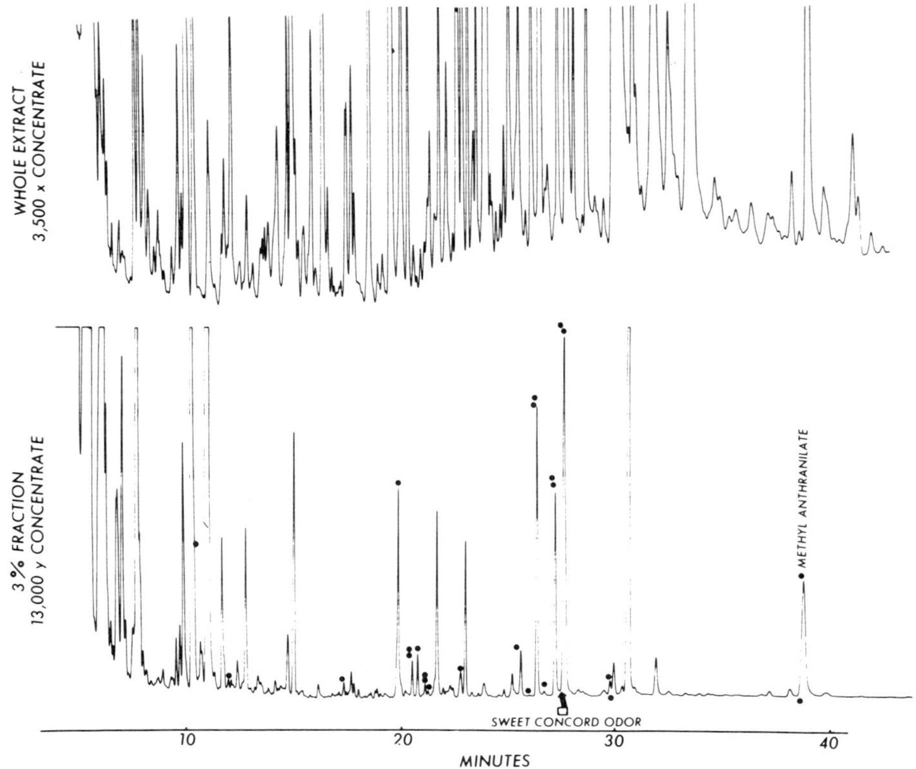

Figure 1. Gas chromatograms from a Freon 113 extract of Concord grape juice concentrated 3,500 times (top) and the 3% ether–pentane fraction of a Florisil separation of the whole Freon 113 extract concentrated 13,000 times (bottom). The dots show regions where odor was detected. Conditions: 30 m $\times$ 0.3 mm carbowax 20 M column, helium gas flow of 23 cm sec^{-1} programmed at 6°C min^{-1}.

grapes at a sufficient concentration to make a contribution to their odor. In fact, preliminary examination showed that the concentration of trans-2-hexen-1-ol in an extract of Niagara was 10 times higher than in a similar Concord extract. The grape breeders of the past have virtually eliminated this trait from commercially important labrusca cultivars. Among the muscadine grapes this may not be the situation, since several of the commercially important muscadine cultivars seem to have a very foxy flavor.

THE FLORAL ODOR

In an attempt to simplify the complex mixture of chemicals present in extracts of labrusca grapes we have used 6% water deactivated Florisil developed with solvents of increasing polarity (15). Made by mixing diethyl ether at 0, 1, 3, 10, 30, and 100% with pentane, each solvent has approximately twice the polarity of the previous one (16). With Concord grape juice extracts the fraction with the most labrusca-like odor character was the 3% fraction which had a very strong sweet floral characteristic. An analysis of this fraction (15) showed the presence of damascenone [1-(2,6,6-trimethyl-1,3-cyclohexadien-1-yl)-2-buten-1-one] a compound previously reported to be present in vinifera grapes(17) as well as other natural products(18).

The possible importance of damascenone to the perception of labrusca character can be inferred from a comparison of the amounts of this compound found in extracts of labrusca and vinifera grape products(15). In a preliminary analysis of products of 11 cultivars, damascenone was highest in extracts of Concord grape juice, 5 ng/g (fresh weight), 37 times higher than that found in an extract of the vinifera cultivar, Pinot chardonnay.

Because damascenone is a ketone its volatility is reduced in products containing sulfur dioxide and thus its contribution to the odor of labrusca wines will be minimized (17). However, its contribution to the odor of grape juice and jelly may be substantial, especially since 5ng/g is 700 times the odor threshold reported for this compound in water (18). In contrast, the methyl anthranilate concentration usually found in Concord grapes is only 50 times its threshold in water(19).

Besides unknown substances yet to be identified, there are other good candidates for the floral odor of labrusca grapes. One such compound is 2-phenylethyl alcohol, which has an odor very reminiscent of labrusca character.

COTTON CANDY

For several years sensory panels (in our laboratory at the Experiment Station at Geneva, New York) have been evaluating the quality of new hybrid grapes in an effort to find a replacement for the labrusca cultivar, Ives. The productivity of this grape has been declining for years because of its sensitivity to air pollution. One descriptor which is frequently associated with the odor of Ives and its hybrid progeny is "cotton candy".

About two years ago a small quantity of a chromatographically pure compound with a strong cotton candy odor was isolated from extracts of Ives grapes. Although there was only enough of this compound to produce one GC-MS run, the spectrum was similar to, though not exactly like, that reported recently by Rapp et al.(5) for a compound identified in Niagara grapes. This compound, called furaneol, (2,5-dimethyl-4-hydroxy-2H-furan-3-one) was described by Rapp as having a strong strawberry odor, whereas Hodge(20) who isolated it from a carbohydrate browning reaction described it as a burnt sugar aroma. As shown in Figure 2 the spectrum we obtained from synthetic furaneol(5,21,22) was superposable on that obtained for the cotton candy component isolated from Ives grapes. Clearly we are describing our sensory perceptions of the same compound using three different descriptors, a situation which, although confusing, is unavoidable at the moment.

Because furaneol can be formed non-enzymatically from carbohydrates it is possible that it is not produced biochemically in natural products either. In fact there is strong evidence that even damascenone is sometimes produced non-enzymatically during the processing of natural products(23,24). It has been demonstrated, however, that furaneol is produced enzymatically from fructose in homogenates of the fruit of arctic bramble berry(25) (*Rubus arcticus* L.) Furthermore, the methyl derivative of furaneol (2,5-dimethyl-4-methyoxy-2H-furan-3-one), also found in labrusca grapes(5) and mango(26), increases rapidly during the ripening of arctic bramble berry to become the major volatile in this fruit . Furthermore, it seems to be the odor-characterizing compound of the arctic bramble berry. We are presently examining the importance of furaneol to the odor of other labrusca cultivars besides Ives,where it seems to cause the cotton candy aroma.

As we learn more about the odor-active components of labrusca grapes we will eliminate such words as floral, foxy, and cotton candy from our sensory vocabulary and replace them with more rigorous descriptors such as damascenone, *trans*-2-hexen-1-ol, and furaneol (Figure 3). But the ultimate test of our understanding of the flavor chemistry of a natural product would be the unambiguous synthesis of a mixture of chemicals with an odor indistinguishable from that of the natural product. Although our present knowledge has yet to pass this test, we are getting closer to an understanding of the chemical causes of labrusca character.

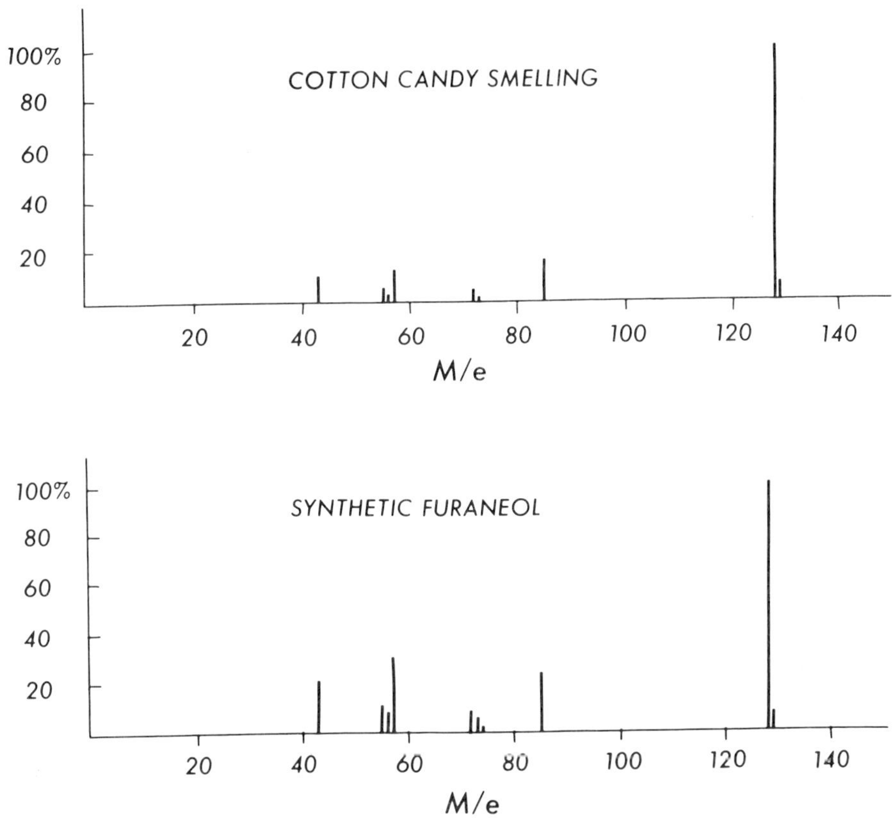

Figure 2. Mass spectra obtained from the cotton candy-smelling compound isolated from Ives grapes (top) and synthetic furaneol (bottom)

COMPOUND	ODOR QUALITY
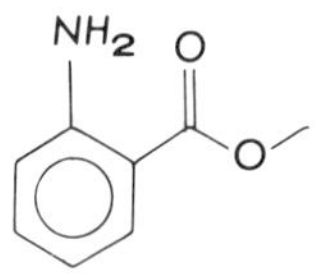 METHYL ANTHRANILATE	METHYL ANTHRANILATE SYNTHETIC GRAPE
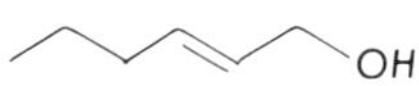 TRANS-2-HEXEN-1-OL	FOXY ANIMAL
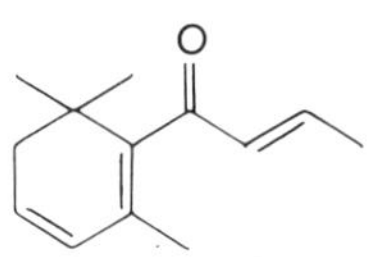 DAMASCENONE	FLORAL FRUITY
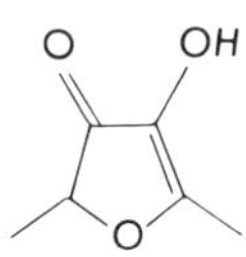 FURANEOL	COTTON CANDY STRAWBERRY BURNT SUGAR

Figure 3. Some compounds isolated from labrusca grapes that may contribute to their odor quality

LITERATURE CITED

1. Galet, Pierre "Cepages et vignobles de France". Montpellier, 1956.
2. Hedrick, U. P. "The Grapes of New York". J. B. Lyon Albany, 1908.
3. Friedman, I. E. Proceedings New York State Hort. Soc. 1976, 121, 131.
4. Nelson, R. R. "Chemical and Sensory Analysis of Wines of Native American Grapes." Thesis, Cornell University, 1977.
5. Rapp, A.; Knipser, W.; Engel, L. L.; Ullemeyer, H.; Heinmann, W. Vitis 1980, 19, 13.
6. Schreier, P. "CRC Critical Reviews in Food Science and Nutrition." 1979, Nov., 59.
7. Powers, F. B.; Chesnut, V. K.; J.A.C.S. 1921, 43, 1741.
8. Gildemeister and Hoffman "Die Aetherischer Oele". 1910, 2 ed., III, 101.
9. Sale, J. W.; Wilson, J. B. J. Agric. Res. 1926, 33, 301.
10. Nelson, R. R.; Acree, T. E.; Lee, C. Y.; Butts, R. M. J. Food Sci. 1977, 42, 57.
11. Robinson, W. B.; Shaulis,N. J.; Pederson, C. S. Fruit Prod. J. Americ. Food Manufacturer. 1949, 29, 36.
12. Military Specification. 1952, MIL-G-3791, 4.
13. Dimick, K. P.; Schultz, T. H.; Makower, B. Food Technol. 1957, 11, 662.
14. Fuleki, T. R. Agric. Res. Inst. Ontario 1977-1978, p. 176.
15. Acree, T. E.; Braell, P.; Butts, R. M. J. Agric. Food Chem. 1981, in press.
16. Snyder, L. R.; Kirkland, J. J. "Introduction to Liquid Chromatography". 1974, John Wiley, New York.
17. Schreier, P.; Drawert, F. Z. Lebens. Unters.-Forsch. 1974, 154, 273.
18. Ohloff, G. Perfumer Flavorists 1978, 3, 13.
19. Fazzalari, F. A. "Compilation of Odor and Taste Threshold Values Data". 1978, ASTM, Philadelphia, 99.
20. Mills, F.; Hodge, J. E. Cardohyd. Res. 1976, 51, 9.
21. Snell, J. M.; McElvan, S. M. Org. Synth. 1943, II, 114.
22. Buchi, G.; Demole, E. J. Org. Chem. 1973, 38, 123.
23. Masuda, M.; Nishimura, K. J. Food Sci. 1980, 45, 396.
24. Schreier, P.; Drawert, F.; Schmid, M. J. Sci. Food Agric. 1978, 29, 728.
25. Kallio, H. J. Food Sci. 1976, 41, 555.
26. Eriksson, C. E. "Progress in Flavour Research". 1979, Applied Science Publishers, London, 159.

RECEIVED April 15, 1981.

3

Quality Characteristics of Muscadine Grape Products

L. F. FLORA[1] and T. O. M. NAKAYAMA

Department of Food Science, University of Georgia Agricultural Experiment Station, Experiment, GA 30212

The muscadine grape vine was discovered in 1584 growing wild near the Scuppernong River located in the Northeast part of North Carolina (1). The muscadine grape embraces two species, Vitis munsoniana and Vitis rotundifolia.

Vitis rotundifolia is the species commonly cultivated commercially in the Southeast. Work with this fruit was begun at the Georgia Experiment Station in 1909 and since that time a large mass of data has been accumulated (2). In most of the region where muscadine grapes are likely to succeed, the temperature does not often go lower than 10°F and it rarely goes to 0°F (3).

Vitis rotundifolia differs so widely botanically from the other species of Vitis that some scientists place it in a separate grouping. This distinctive grouping has become so well known in the southern states that to the grower there are two classes of grapes, muscadine or Scuppernong and the so-called bunch grapes (4).

The muscadine group of grapes includes both the light or bronze varieties and the dark or black varieties. The most widely known variety of this group is the Scuppernong. The whole group is sometimes referred to as Scuppernong; however, the Scuppernong is only one of the white varieties belonging to the muscadine group (5).

The fruit clusters of the muscadine group are small with the berries in many varieties tending to shell or fall from the cluster when ripe. The fruit is very characteristic having a thick leathery appearing skin with small russet dots in most varieties. The individual berries are large and round. The pulp is quite tough, but this toughness is not localized around the seed as with the labrusca. The fruit has a characteristic musky odor and in many varieties tends to ripen unevenly in the cluster.

In contrast, the fruit cluster of the Vitis labrusca

[1] Current address: American Home Foods, Milton, PA 17847.

0097-6156/81/0170-0021$05.00/0

varieties are larger and usually tapering with the berries ripening evenly and strongly holding to the cluster. The berries are usually covered with a waxy secretion known as bloom, which is not present in the rotundifolia grape.

Both species of grapes, labrusca and rotundifolia, can be classified as slip skins, that is the skin is not adherent to the flesh (4). The mucilagenous pulp and leathery skin of crushed muscadines are difficult to handle in wine making operations and the recovery of clear juice is low. Sugar percentage of cultivated muscadines is too low for wine making without amelioration of the must (6).

In a 1934 study reported by Armstrong, Pickett, and Murphy (5), studying 27 varieties of muscadine grapes, they found the number of seeds per berry ranged from 1.8 to 3.4, the percentage of seed was 2.1 to 6.0, the percentage of skin and pulp ranged from 15.9 to 30.7 and the percentage of expressed juice ranged from 64.6 to 80.2. The chemical characteristics of the expressed juice from 43 varieties of grapes taken during that 1934 season were total solids ranging from 12.82% to 21.32%, pH ranging from 3.42 to 2.96, titratable acidities 54 to 174 cc of 0.1N NaOH per 100 cc of juice and the tannin content 20 to 417 mgs per liter. Flora (7) updated these figures in a 1977a report.

Savage (4) also reported in his 1941 report that practically all of the sugars found in the Concord grape were reducing sugars while in the muscadine variety such as the Hunt, Stuckey and Dulcet, a large proportion of the sugar was sucrose. In general, the Concord juice has considerably more titratable acidity than the muscadine juices. The muscadine juices also had considerably lower tannin contents than Concord juices.

Carroll, Hoover and Nesbitt (8), in a 1971 study, also recognized that the level of total soluble solids in the mature fruits of *Vitis rotundifolia* were low, ranging from 10 to 18%, while ripe *Vitis vinifera* and *Vitis labrusca* berries generally ranged between 19 to 24% and 13 to 21%, respectively. Glucose and fructose are known to be the principal sugars of *Vitis vinifera* with sucrose present in only trace quantities, whereas up to 5.2% of sucrose was found in the 12 cultivars of *Vitis rotundifolia* studied by Carroll (8).

Research has shown that ripeness is important to the quality of juices and wines made from muscadines. Lanier and Morris (9) separated muscadine grapes into five density grades using four brine solutions and reported that the flavor quality of juice from muscadines sorted into ripeness levels by the density flotation method increased with the ripeness or soluble solids. Carroll (10) reported that the quality or acceptability of wine made from grapes separated into ripeness levels by light sorting increased to a point and then dropped off as the grapes became overripe.

Carroll (11) reported in 1975 that muscadine cultivars lacked sufficient sugar to make balanced and stable wines. Lack of sufficient sugar is a common occurrence for grapes growing in the eastern United States; therefore, sugar or sucrose is added to all musts in order to increase the sugar contents to 21° Brix. Apparently the hulls of muscadines contain appreciable amounts of acid which are released into the fermenting wine. After fermentation on the skin and pressing, most musts were high enough in acidity to require amelioration with 21° Brix sugar syrup in order to reduce the total acidity of the finished wine to a palatable level.

One of the unique aspects of muscadine grapes is their complement of anthocyanin pigments. In a 1940 report, Brown (12) stated that the anthocyanin of the Hunt muscadine grape is probably a 3,5-diglucoside of 3-O methyldelphinidin. This pigment was commonly referred to as muscadinin. Ribereau-Gayon (13) showed that the muscadine pigment complex is composed of five different anthocyanins: delphinidin, petunidin, cyanidin, malvidin, and peonidin. He also demonstrated that they occur only as diglucoside linkages at positions 3 and 5. Ballinger, Maness and Nesbitt (14) confirmed these findings with seven cultivars of muscadines and three advanced selections from the North Carolina Grape Breeding Program. All contained the same sugar, the 3,5-diglucosides and the five anthocyanins as previously reported by Ribereau-Gayon (13).

Further studies by Nesbitt, Maness, Ballinger, and Carroll (15) indicated the potential color quality of muscadine grapes can be effectively determined without going through the wine making process. They demonstrated that good wine color was directly related to the presence of large amounts of malvidin-3,5,diglucoside and to a lesser degree to the petunidin-3,5-diglucoside. Delphinidin-3,5-diglucoside was the most abundant form of all the different anthocyanins present; however, the researchers found no apparent correlation between the quantity of the delphinidin-3,5-diglucoside present in the grapes and good wine color quality. Based on these studies the researchers concluded that increased methylation of individual anthocyanins resulted in greater pigment stability with the malvidin-3,5-diglucoside having the most desirable color effect at a concentration of eight or more milligrams per 100 grams of fresh grapes.

Ballinger (16) in a subsequent 1974 study also confirmed these results with wines. He reported that the glycosides of malvidin are most stable in wines followed by peonidin, petunidin, cyanidin and delphinidin in decreasing order of stability.

Flora (17) in 1978 reported a range in anthocyanin concentrations from 40 to 403 mgs per 100 g of fruit in 11 cultivars of muscadines with wide variations in the relative contents of individual anthocyanins. Delphinidin was usually the most prevalent anthocyanin in the pigment complex followed by cyanidin or petunidin, peonidin and malvidin. Cultivars with relatively high malvidin and total anthocyanin contents yielded juices and jellies with the highest quality and most stable colors. Harvest date influenced relative percentages of peonidin, malvidin and delphinidin. Maturity level influenced the relative percentage of malvidin and concentrations of all anthocyanins.

One of the more serious problems encountered with muscadine juices and wines is in the preservation of the desirable appearance or color of the product. Prevention of undue discoloration is quite difficult. Few studies have been conducted to explain the mechanisms of browning and to design appropriate methods for the elimination or the partial prevention of this phenomenon. The tendency of the anthocyanin-3,5-diglucosides which are present in the muscadine grapes to brown more severely than the monoglucosides has been established by Robinson (18) and Van Buren (19). Though oxidation appears to play a part in the browning of the muscadine pigments, pigment polymerization or condensation also seems to be involved.

Though Woodroof (20) found that adding ascorbic acid to containers of pasteurized muscadine grape juice which had been opened aided in reducing darkening and flavor loss due to oxidation, Flora (21) found that ascorbic acid added to containers of red muscadine grape juice accelerated the fading and browning of the juice in storage. Refrigeration was the single most effective tool in maintaining color and flavor of juice. Sulfur dioxide improved retention of juice color.

In a 1976 study by Flora (22) reporting the time-temperature influence on muscadine grape juice quality, he found that juice color became unacceptable as the browning index, the ratio of absorbance at 520 nm over the ratio of absorbance at 420 nm, dropped below 3.4 with the tristimulus A value dropped below +22. The color damage from heating the red muscadine juices progressed in a roughly logarithmic fashion from 46°C to 121°C with a Q_{10} rate change factor of approximately 1.4. Temperature residence and total heat input had the greatest effect in color changes. Aroma changes preceded unacceptable color changes in the heated juices.

Muscadines of the rotundifolia variety have been used to furnish the characteristic aroma to blends predominating in the neutral flavored viniferas. In fact, a hybrid containing the desirable aroma characteristic and disease resistance of the rotundifolia with the vigor and crop producing traits of some of the vinifera varieties would be highly desirable.

The rose-like aroma of ß-phenylethanol is an important constituent of the *Vitis rotundifolia* aroma. Bitter phenolics present in *Vitis labrusca*-*Vitis vinifera* hybrids are especially important in hybrids with *Vitis cinerea* and *Vitis cortfolia* but are not present in Vitis rotundifolia. Olmo (6), reporting in 1971 on hybrid trials with Vitis vinifera and Vitis rotundifolia, found that the aromatic components of the Vitis rotundifolia in contrast with those of the Vitis labrusca are seldom recovered in hybrids with Vitis vinifera or are present in such diluted form as to be unrecognizable.

In a 1956 report, Kepner and Webb (23) reported that the presence of ß-phenylethanol with its rose-like odor undoubtedly is important in the aroma of the rotundifolia grape. The authors identified 18 components present in muscadine grape flavor including ethyl alcohol, butyl alcohol, hexyl alcohol, an acetate ester, a laurate ester, methyl alcohol, hexanal, 2-hexenal, isoamyl alcohol, acetaldehyde, isobutyraldehyde, biacetyl, ethyl acetate, a caproate ester, a caprylate ester, a caprate ester, a laurate ester, and they also identified as probably being present methyl ethyl ketone, in addition to the ß-phenylethanol. Of particular importance is that the preliminary test from this report showed the absence of nitrogen-, halogen-, and sulfur-containing compounds in the rotundifolia volatile essence. The lack of the nitrogen-containing compound of volatile esters in rotundifolia grapes may be a means of differentiating it from labrusca grapes which possess nitrogen-containing anthranilic esters.

The more recent studies, Flora et al. (24), identified several constituents of muscadine grape essence using gas-liquid chromatography in conjunction with mass spectrometry and infrared spectrometry. Components identified included methanol, ethanol, n-butanol, 2-methyl-1-butanol, n-hexanal, trans-2-hexene-1-al, ß-phenylethanol, acetaldehyde, ethyl acetate, ethyl propionate, propyl acetate, butyl acetate, ethyl caprate, benzyl acetate, toluene, m-xylene, and d-limonene. The authors assessed the relative importance of these components using gas chromatographic effluent sniffing. ß-phenylethanol seems to be the component most like the muscadine flavor with benzyl acetate and one or two unknowns also contributing a grapey flavor or aroma.

In a subsequent study by Sistrunk (25), it was found that the concentrations of some of the volatiles which characterized the good muscadine flavor have a curvilinear relationship with acceptability. The most acceptable cultivars of muscadine grape juice were those which contained intermediate levels of important volatile components. In general, low and medium boiling volatiles were decreased by hot-pressing of muscadine grape juice. Flavor quality of hot-pressed juice seemed to be determined more by nonvolatile components than by volatile flavor components. Storage temperature had some effect upon volatiles concentration in processed muscadine grape juice. A few volatiles including ethyl acetate, ethyl propionate and butyl acetate were significantly lower in juice stored at 21°C as opposed to juice stored at a refrigerated temperature of

4°C. The overall acceptability of juice stored at 21°C was significantly lower than the juice stored at 4°C. Volatiles concentrations significantly decreased during 12 weeks of storage, probably due to oxidation and general degradation. The most pronounced decreases occurred early in the storage period. Using objective and subjective techniques, Sistrunk (25) determined that ethyl acetate, benzyl acetate and ß-phenylethanol are probably of primary importance to the typical flavor of muscadines. Ethyl acetate comprised 25% 91% of the total volatile concentrations of juices and at optimum levels was responsible for fruity aroma and flavor. Benzyl acetate and ß-phenylethanol are the components which are most associated with the muscadine grape aroma and flavor.

Unknown components detected in essence extracts from muscadine grape juice by effluent sniffing also seemed to be of primary importance to the flavor of muscadine grapes. Efforts are continuing to identify these components.

Literature Cited

1. Reimer, F. C. 1909. North Carolina Exp. Sta. Bull. 201. p. 21.
2. Stuckey, H. P. 1919. Ga. Expt. Sta. Bull. No. 113. p. 62-74.
3. Dearing, C. 1947. Farmer's Bull. No. 1785, U.S. Dept. of Agriculture pp. 1-29.
4. Savage E. F. 1941. Ga. Expt. Sta. Bull. 217. p. 36
5. Armstrong, W. D., Pickett, T. A., and Murphy, M. M., Jr. 1934. Ga. Expt. Sta. Bull. 185. p. 1-35.
6. Olmo, H. 1971. Am. J. Enol. Viti. 22:87-91.
7. Flora, L. F. 1977a. J. Food Sci. 42:935-938, 952.
8. Carroll, D. E., Hoover, M. W., and Nesbitt, W. B. 1971. J. Am. Soc. Hort. Sci. 96:737-740.
9. Lanier, M. R. and Morris, J. R. 1979. J. Am. Soc. Hort. Sci. 104:249-252.
10. Carroll, D. E. Ballinger, W. E., McClure, W. F., and Nesbitt, W. B. 1978. Am. J. Enol. Vitic. 29(3):169-171.
11. Carroll, D. E., Nesbitt, W. B., and Hoover, M. W. 1975. J. Food Sci. 40:919-920.
12. Brown, W. L. 1940. J. Am. Chem. Soc. 62:2808-2810.
13. Ribereau-Gayon, P. 1964. An. Physiol. Veg. 6(3):211.
14. Ballinger, W. E., Maness, E. P., Nesbitt, W. B., and Carroll, D. E. 1973. J. Food Sci. 38:909-190.
15. Nesbitt, W. B., Maness,E. P., Ballinger, W. E., and Carroll, D. E. 1974. Am. J. Enol. Vitic. 25:30-32.
16. Ballinger, W. E., Maness, E. P., Nesbitt, W. B., Makus, D. J., and Carroll, D. E. 1974. J. Am. Soc. Hort. Sci. 99:338-341.
17. Flora, L. F. 1978. J. Food Sci. 43:1819-1821.

18. Robinson, W. B., Weirs, L. D., Bertino, J. J., and Mattick, L. R. 1966. Am. J. Enol. Vitic. 17:178-184.
19. Van Buren, J. P., Bertino, J. J., and Robinson, W. B. 1968. Am. J. Enol. Vitic. 19:147-154.
20. Woodroof, J. G., Cecil, S. R., and DuPree, W. E. 1956. Ga. Agric. Exp. Sta. Bull. N.S. 17, 35 pp.
21. Flora, L. F. 1977b. Am. J. Enol. Vitic. 28(3):171-175.
22. Flora, L. F. 1976. J. Food Sci. 41:1312-1315.
23. Kepner, R. E. and Webb, A. D. 1956. Am. J. Enol. Vitic. 7:8-18.
24. Flora, L. F., Coleman, R. L., and Bryan, W. L. 1977. Ann. Meeting Inst. Food Technol. p. 108
25. Sistrunk, C. P. 1979. M.S. Thesis, U. of Ga. 92 p.

RECEIVED April 27, 1981.

4

Raisin and Dried Fig Volatile Components: Possible Insect Attractants

RON G. BUTTERY, RICHARD M. SEIFERT and LOUISA C. LING
Western Regional Research Laboratory, SEA, USDA, Berkeley CA 94710

EDWIN I. SODERSTROM and ALBERT P. YERINGTON
Stored Products Insects Research Laboratory, SEA, USDA, Fresno, CA 93727

The vacuum steam volatile oils of both raisins and dried figs have been analyzed by capillary GLC-mass spectrometry. A total of 38 components were identified in the volatile oil of raisins and 34 components in the volatile oil of dried figs. Major volatile components in both raisins and dried figs included fatty acid degradation products such as nonanoic and octanoic acids, (*E*)-2-decenal, (*E*)-2-octenal and nonanal. Benzaldehyde was an additional major component of dried figs. The most unusual components were 2-hexyl-3-methyl-maleic anhydride and 1-octen-3-one in the raisin volatile oil.

Certain insects such as the Indian meal moth (*Plodia interpunctella*) and saw tooth grain beetle (*Oryzaephilus surinamensis*) infest dried fruits. It seems reasonable that these insects are attracted to the dried fruit, at least to some extent, by the characteristic odor of the dried fruit. The present study was begun to identify the volatile (odor) components of raisins and dried figs in order to be in a position to test these compounds for attraction. The volatiles of raisins had been previously studied by Ramshaw and Hardy (1). The volatiles of fresh figs had been previously studied by Jennings (2).

EXPERIMENTAL

Materials.

Good quality dried raisins (dried Thompson seedless grapes), dried figs and fresh figs (Calimyrna variety) were obtained from processors in the Fresno, California area and from local markets for comparison. Authentic chemical compounds for comparison were obtained from reliable commercial sources or synthesized by known methods.

Isolation of Volatile Oils.

Whole raisins (1500g) were placed in a 12 L flask together with 6 L of odor free water. A Likens-Nickerson steam distillation continuous extraction head was attached to the flask.

Purified hexane (100 mL) was placed in a 250 mL flask attached to the solvent arm of the head. The isolation was carried out under reduced pressure (100 mm Hg) for 3 hours with the raisins at ca. 50°C. Cooling of the head condenser was carried out with an ethanol-water mixture at 0°C. A dry ice cooled reflux condenser was placed on the outlet of the head. After the isolation the hexane extract was dried by freezing out any drops of water and filtered. The hexane solution was then concentrated using low hold up distillation columns to give the raisin volatile oil.

The volatile oil from whole dried figs was isolated using essentially the same procedure as for the raisins except that the weight of dried figs was 1 Kg.

The volatile oil from whole fresh figs was isolated also using essentially the same procedure as for raisins except using 3 Kg of fresh figs.

Volatile Oil from Fermented Figs.

Dried figs (1 Kg) were added to water (1.5 L) in a 5 L beaker. Bakers yeast (0.25 g) was then added and the mixture stirred briefly. The beaker was covered with cheesecloth and the mixture allowed to ferment at room temperature (25°C) for 3 days. The mixture together with 4 L of odor free water was treated using a Likens-Nickerson head vacuum steam distillation continual extraction procedure as described above for raisins.

Other batches of fermented fig prepared as above were extracted directly with freshly distilled diethyl ether (3 x 500 mL). The ether was then concentrated to give the volatile oil as for the hexane extracts.

Direct Vapor Analyses.

Samples of fresh fig or rehydrated dry fig (100 g) were enclosed in a wide mouth Erlenmeyer flask (250 mL) and the flask covered with aluminum foil. Samples of vapor (10 mL) from above the figs was drawn up into a clean dry glass syringe and injected directly into the capillary GLC column (cf. 3). This procedure was used for both the capillary GLC-mass spectrometry analyses and the quantitative analyses. Quantitative analyses were made by comparison of peak areas with those from vapor samples above known concentrations of the compounds in water solution.

Capillary GLC-Mass Spectral (GLC-MS) Analysis.

This was carried out using a 150 m long by 0.64 mm i.d. Pyrex glass capillary column coated with Carbowax 20-M. The column was held at 50°C for 30 minutes after injection and then temperature programmed at 1°/minute to 170°C and held at this upper limit. A single stage Llewellyn-Littlejohn silicone rubber membrane separator was used to couple the end of the capillary column to the mass spectrometer (a modified Consolidated 21-620 cycloidal type). Electron ionization was 70 eV.

RESULTS AND DISCUSSION

Raisin Volatiles.

Steam distillation continuous extraction of the raisins under reduced pressure gave a volatile raisin oil which amounted to 5-10 ppm of the raisins. The oil was analyzed by capillary GLC-mass spectrometry (GLC-MS). The analysis was repeated with volatile oils from several different samples of raisins. Figure I shows a GLC analysis of typical oil. The results are listed in Table I. Peak numbers corresponding to the peaks in Figure I are shown in the first column of Table I alongside the compound's name. Some idea of the approximate relative percentages of the components in a typical oil (calculated from peak areas) is also listed in Table I.

The major compounds in the raisin volatile oil include the aliphatic acids octanoic, nonanoic, hexanoic, heptanoic, and decanoic acids as well as 2-hexyl-3-methylmaleic anhydride (previously identified by some of the authors, 4), nonanal, phenylacetaldehyde and 2-pentylfuran. The most unusual component is 2-hexyl-3-methylmaleic anhydride. Anhydrides are usually hydrolyzed by water and are not likely to occur in foods because most foods have high concentrations of water. Conditions in dried foods are apparently, however, suitable for the stability and possibly the formation of anhydrides. It is surprising that this anhydride survives the vacuum steam distillation isolation procedure although the relatively low temperature (ca. 50°C) and likely rapid transfer to the hexane solvent are probably favorable.

Another unusual component is the potent odorant 1-octen-3-one which has been reported to have a mushroom-metallic aroma and has been found in cooked mushroom (5) and artichoke (6).

The major components, except for phenylacetaldehyde and possibly 2-hexyl-3-methylmaleic anhydride, seem to be derived from oxidative lipid breakdown in common with the volatiles of many other foods and plant materials.

Eleven of the compounds in Table I had been identified previously in raisins by Ramshaw and Hardy (1). These previously identified compounds have been noted on Table I (see footnote d). The relative quantitative pattern found in the present work seems to be quite different from that found by Ramshaw and Hardy. This difference may be due to the method and degree of drying used. Ramshaw and Hardy analyzed Australian raisins that had been dried to 10% moisture whereas California raisins are dried to ca. 14%. The Australian raisins

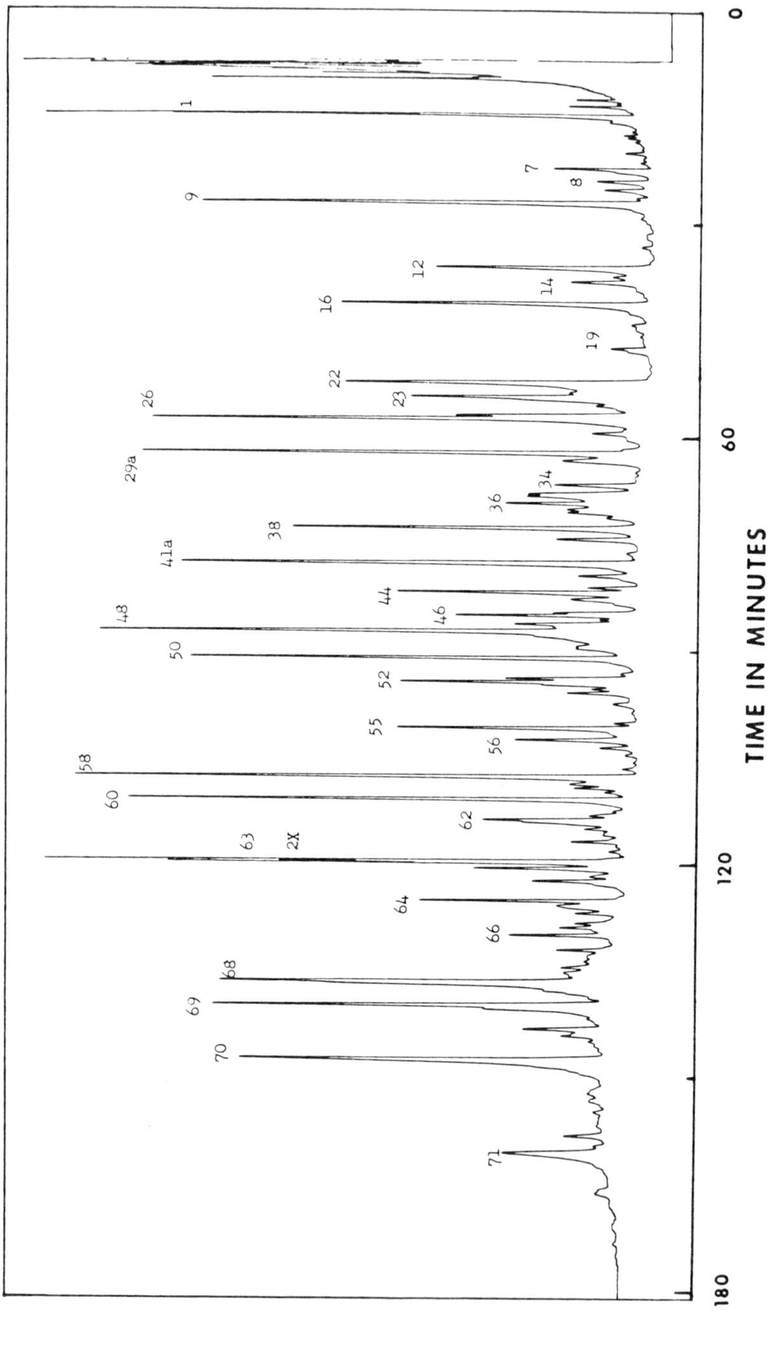

Figure 1. Capillary GLC analysis of the vacuum steam volatile oil of raisins. GLC conditions are described in text.

TABLE I

Compounds Identified in the Vacuum Steam Volatile Oil of Raisins.

Peak No. Fig. I	Compound[a]	Characteristic Mass Spectral Ions[b]	Kovat's GLC Index	Rel. %
Alkanals				
1	Hexanal[d]	44, 56, 72, 82, 67, 100	1108	1.5
7	Heptanal	44, 70, 81, 86, 96, 114	1190	0.3
12	Octanal	43, 44, 84, 100, 110, 128	1290	0.2
22	Nonanal	57, 44, 98, 82, 114, 142	1390	5.4
36	Decanal	57, 44, 82, 112, 128, 156	1500	1.2
Alkenals				
6	(E)-2-Hexenal[d]	41, 42, 55, 69, 83, 98	1230	0.1
16	(E)-2-Heptenal[d]	41, 55, 83, 70, 68, 112	1330	2.1
26	(E)-2-Octenal	41, 55, 70, 83, 97, 126	1430	2.7
38	(E)-2-Nonenal	41, 70, 83, 96, 111, 140	1530	1.7
48	(E)-2-Decenal	43, 70, 83, 97, 110, 154	1630	2.0
55	(E)-2-Undecenal	41, 70, 83, 97, 121, 168	1740	1.2
Alkadienals				
34	(E,E)-2,4-Heptadienal	81, 39, 53, 110, 67, 95	1480	1.7
52	(E,E)-2,4-Nonadienal	81, 41, 67, 95, 138, 109	1660	0.4
56	(E,Z)-2,4-Decadienal	81, 41, 67, 95, 152, 123	1740	0.2
58	(E,E)-2,4-Decadienal	81, 41, 67, 95, 152, 123	1790	2.0
Alkenones				
14	1-Octen-3-one	55, 70, 43, 97, 83, 111	1290	0.2
Alkanols and Alkenols				
29a	1-Octen-3-ol	57, 72, 85, 81, 99, 110	1420	2.6
41a	Octanol	56, 42, 70, 31, 84, 112	1530	1.7
50	Nonanol	56, 70, 42, 31, 98, 126	1630	1.6
57	Decanol	70, 42, 83, 31, 97, 112	1740	0.3
Alkanoic Acids				
61	Hexanoic acid	60, 73, 57, 45, 55, 87	1880	0.5-4.4
64	Heptanoic acid	60, 73, 55, 45, 87, 101	1990	0.5-4.0
68	Octanoic acid	60, 73, 45, 85, 101, 115	2100	4-8.4
70	Nonanoic acid	60, 73, 45, 115, 129, 158	2220	3-5.5
71	Decanoic acid	60, 73, 57, 129, 87, 172	2330	1-1.9

Table I continued

Terpenoids				
43	(E)-2-Methyl-2,4-heptadien-6-one	109, 43, 81, 53, 124, 79	1590	0.1
52	alpha-Terpineol[d]	59, 93, 81, 121, 136, 139	1710	0.4
62	Geranylacetone	43, 69, 93, 136, 151, 194	1850	1.4
Benzene, Naphthalene and Furan Compounds				
9	2-Pentylfuran	81, 53, 138, 39, 95, 68	1240	3.4
29	Furfural[d]	39, 96, 95, 67, 42, 50	1450	0.5
35	2-Acetylfuran[d]	95, 110, 39, 68, 53, 51	1490	1.0
36	Benzaldehyde[d]	77, 105, 106, 51, 50, 39	1520	0.8
41,	5-Methylfurfural[d]	110, 109, 53, 39, 81, 95	1560	2.2
47	Phenylacetaldehyde[d]	91, 92, 120, 65, 39, 51	1650	4.0
53	Naphthalene	128, 64, 51, 63, 102, 77	1690	0.4-2.0
59	2-Methylnaphthalene	142, 141, 115, 71, 57.5, 63	1800	0.3
65	Dimethylnaphthalene	156, 141, 155, 115, 128, 76	2000	0.2
Others				
69	2-Hexyl-3-methylmaleic anhydride	126, 43, 98, 67, 140, 196	2090	5.0
44	N-Ethyl-2-formylpyrrole[e]	123, 94, 108, 39, 66	1600	1.0

[a]Mass spectrum (complete spectrum) and Kovat's GLC retention index of all compounds listed are consistent with that of authentic samples, except for e.

[b]Not necessarily the most intense ions but 5 of those considered the most characteristic for that compound. Ions in descending order of intensity with the most intense ion first and molecular ion underlined.

[c]Kovat's index for the Carbowax 20-M coated Pyrex capillary described in the experimental section.

[d]Previously identified in raisins by Ramshaw and Hardy, 1969.

[e]No authentic sample available. Mass spectrum consistent with published data (8).

had showed a much greater relative amount of sugar degradation products such as furfural and biacetyl.

It is interesting that Ramshaw and Hardy (1) identified a methylformylpyrrole whereas 1-ethyl-2-formylpyrrole was found in the present work.

Fig Volatiles.

The vacuum steam volatile oil of dried fig amounted to 5 ppm of the dried fig. Volatile oils from several different samples of dried figs were isolated and analyzed by GLC-MS. Figure II shows a GLC analysis of a typical dried fig volatile oil. Table II lists the components identified and compares the amounts found with those found in the present work in fresh figs of the same variety. Mass spectral and GLC retention data of compounds already listed in Table I are not repeated. Data for additional compounds have been included in Table IIA.

The ethanol, ethyl acetate, methyl acetate and acetaldehyde had been found in fresh figs by previous workers (2). These are by far the most abundant volatile components in the fresh fig. In the present work these very volatile components were analyzed (GLC-MS) using direct injection of the vapor above the figs onto the glass capillary column. As much as 0.5% ethanol was found in some samples of good quality fresh ripe figs. There was considerable variation, however. Other esters reported in the previous study (2) were not detected using the methods of the present study. The ethanol, ethyl and methyl acetates and acetaldehyde might be formed by fermentation occurring in the fresh fig. This fermentation is not apparent from the appearance and flavor of the fig which appear quite normal and "fresh".

The very volatile fermentation components discussed above seem to be largely lost in the dry fig. The major component of the dried fig volatile oil is benzaldehyde. Other major components include the aliphatic acids octanoic and nonanoic acids. These acids were also major components of raisins and have aromas reminiscent of both raisins and dried figs. They had previously also been found as major components of dried almond hulls (7) which is essentially also a dried fruit. The 2-hexy-3-methylmaleic anhydride common to raisins and dried almond hulls (4) was not detected in dried figs.

Unusual components found in the dried fig volatile oil include 3-phenylpropanal (dihydrocinnamaldehyde), linalool oxide A (2-methyl-2-vinyl-5(2'-hydroxy-2'propyl)-tetrahydrofuran), linalool oxide C (5-hydroxy-2,6,6-trimethyl-2-vinyltetrahydropyran) and N-ethyl-2-formylpyrolle. The last compound was also identified in raisin volatiles but in both foods the identification was only by comparison with a published spectrum (8) and no authentic sample was available.

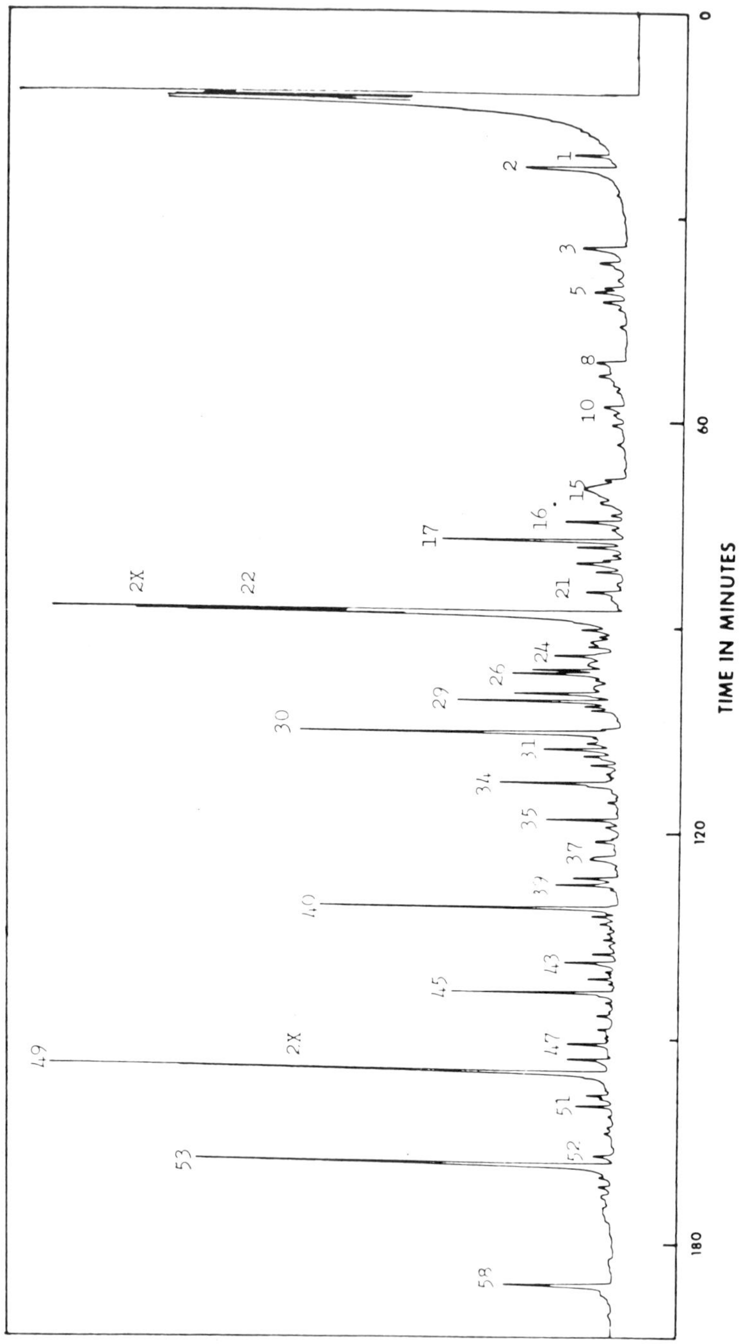

Figure 2. Capillary GLC analysis of the steam volatile oil of dried figs. GLC conditions are described in text.

TABLE II

Volatile Components Identified in Dried and Fresh Figs.

Peak No. (Fig. II)	Compound[a]	Concentration in fresh or dried dried in fig in ppm	
		Dried Fig	Fresh Fig
-	Acetaldehyde	<1 ppm	7-40
-	Methanol	<1	5
-	Ethanol	<10	144
-	Methyl acetate	<1	4
-	Ethyl acetate	<1	9
-	Vacuum steam volatile oil	5	3
	Alkanals		
2	Hexanal	0.3	0.04
3	Heptanal	0.1	0.03
8	Octanal	0.1	0.01
15	Nonanal	0.2	0.08
	Alkenals		
-	(E)-2-Hexenal	-	0.005
10	(E)-2-Heptenal	0.1	0.03
16	(E)-2-Octenal	0.2	0.06
23	(E)-2-Nonenal	0.05	0.03
30	(E)-2-Decenal	0.1	0.2
36	(E)-2-Undecenal	0.04	0.04
	Alkadienals		
20	(E,Z)-2,4-Heptadienal	0.05	0.1
21	(E,E)-2,4-Heptadienal	0.1	0.1
-	(E,E)-2,4-Nonadienal	-	0.03
-	(E,Z)-2,4-Decadienal	-	0.06
38	(E,E)-2,4-Decadienal	0.15	0.2
	Alkanones and Alkenones		
2a	2-Hexanone	0.05	-
3a	2-Heptanone	0.03	-
8a	2-Octanone	0.03	-
22a	(E,Z)-3,5-Octadien-2-one	0.03	-
25	(E,E)-3,5-Octadien-2-one	0.1	0.02
	Alkanols and Alkenols		
-	Hexanol	-	0.05
-	(Z)-3-Hexenol	-	0.01
-	Heptanol	-	0.005

TABLE II continued

Peak No. Fig. II	Compound	Dried Fig	Fresh Fig
19	1-Octen-3-ol	0.1	-
24a	Octanol	0.03	0.02
33	Nonanol	0.02	0.08
	Alkanoic Acids		
40	Hexanoic acid	0.05	-
45	Heptanoic acid	0.04	-
49	Octanoic acid	0.2	-
53	Nonanoic acid	0.15	-
58	Decanoic acid	0.1	-
	Aliphatic Esters		
6	Ethyl hexanoate	0.05	0.02
	Terpenoids		
-	Limonene	-	0.005
12	2-Methyl-2-hepten-6-one	-	0.01
18	Linalool oxide (A)	0.1	-
24	Linalool	0.02	-
28	(E)-2-Methyl-2,4-heptadien-6-one	-	0.09
35	Linalool oxide (c)	0.04	-
41	Geranylacetone	0.05	0.04
	Benzene, Furan and Heterocyclic Compounds		
5	2-Pentylfuran	0.05	0.01
22	Benzaldehyde	1.8	0.3
29	N-Ethyl-2-formylpyrolle[e]	0.2	-
37	3-Phenylpropanal	0.05	-
-	Ethyl phenylacetate	-	0.05
-	2-Phenylethanol	-	0.06

[a]Mass spectrum (complete spectrum) and Kovat's GLC retention index of all compounds listed are consistent with that of authentic samples. Except for e, as for Figure I.

TABLE IIA

Mass spectral and GLC data on fig and dried fig volatiles (not already listed in Table I).

Peak No. (Fig. II)	Compound[a]	Characteristic Mass Spectral Ions[b]	Kovat's GLC Index[c]
Alkanals			
-	Acetaldehyde	29, 44, 43, 42	690
Alkadienals			
-	(E,Z)-2,4-Heptadienal	81, 39, 53, 110, 67, 95	1450
Alkanones and Alkenones			
2a	Hexan-2-one	43, 58, 57, 100, 71, 85	1100
3a	Heptan-2-one	43, 58, 71, 59, 114, 99	1190
8a	Octan-2-one	43, 58, 71, 59, 128, 85	1290
22a	(E,Z)-3,5-Octadien-2-one	43, 95, 81, 124, 53, 109	1525
25	(E,E)-3,5-Octadien-2-one	43, 95, 81, 124, 53, 109	1550
Alkanols			
-	Methanol	31, 32, 29	870
-	Ethanol	31, 45, 46, 29, 43	920
Terpenoids			
-	Limonene	68, 93, 136, 79, 107, 121	1180
-	Linalool Oxide A [2-Methyl-2-vinyl-5--(2'-hydroxy-2'-propyl)--tetrahydrofuran]	59, 43, 94, 111, 155	1470
35	Linalool oxide C [5-hydroxy-2,6,6--trimethy-2-vinyl--tetrahydropyran]	68, 59, 43, 94, 112, 155	1725
	Aromatic		
-	Ethyl phenylacetate	91, 164, 65, 39, 51, 119	1770
-	2-Phenylethanol	91, 92, 122, 65, 39, 51	1890
37	3-Phenylpropanal	91, 92, 134, 78, 105, 51	1760

TABLE IIA continued

Aliphatic Esters		
Methyl acetate	43, 74, 42, 59, 29	780
Ethyl acetate	43, 29, 45, 61, 70, 88	870
Ethyl hexanoate	43, 88, 99, 60, 73, 101	1230

a,b,c, As for Table 1.

Insect Tests With Raisin and Dried Fig Volatiles.

Some tests have been begun with the raisin and dried fig volatiles in their ability to attract the Indian meal moth (*Plodia interpunctella*) and sawtoothed grain beetle (*Oryzaephilus surinamensis*) but no reportable results have been obtained yet.

Fermented Figs and Drosophila Attraction.

Nitidulids and *Drosophila* spp. are attracted to figs during the ripening phase in the orchard. Experiments (9), have shown that re-hydrated dried figs allowed to ferment for ca. 3 days were the most attractive to these insects. Studies were carried out in the present work to identify the components present in the fermenting rehydrated fig and to test these components against *Drosophila*. The volatiles found in the fermented re-hydrated dry fig were similar to those found in other fermented products (cf. 10). There was considerable variation with different fermentation lots but in general concentrations were of the order of the following: 0.4-2% ethanol, 40 ppm acetaldehyde, 10 ppm ethyl acetate, 20 ppm 2-phenylethanol, 1-2 ppm 2-phenylethyl acetate, 2.5 ppm 3-methylbutyl acetate, 30 ppm 2-methylpropanol, 100 ppm 3-methylbutanol, 4 ppm ethyl hexanoate, 8 ppm ethyl octanoate, 6 ppm ethyl decanoate, 1 ppm ethyl dodecanoate and concentrations roughly of the order of 1-10 ppm for the free acids octanoic, decanoic, undecanoic and dodecanoic. An undetermined amount of acetic acid was also identified. All of the compounds were tested against *Drosophila* using a laboratory olfactometer. The only compounds to show some attractivity were ethanol, ethyl acetate, acetaldehyde, acetic acid (as a mixture, cf. 9, 11) also 2-phenylethanol, 2-phenylethyl acetate and 3-methylbutyl acetate. Further studies are in progress with varying mixtures of these compounds.

LITERATURE CITED

1. Ramshaw, E. H.; Hardy, P. J. J. Sci. Fd. Agric., 1969, 20, 619.
2. Jennings, W. Food Chem. 1977, 2, 185.
3. Buttery, R. G.; Teranishi, R. J. Agric. Food Chem. 1963, 11, 504.
4. Buttery, R. G.; Seifert, R. M.; Haddon, W. F.; Lundin, R. E. J. Agric. Food Chem., 1980, 28, 1336.
5. Cronin, D. A.; Ward, M. K. J. Sci. Fd. Agric., 1971, 22, 477.
6. Buttery, R. G.; Guadagni, D. G.; Ling, L. C. J. Agric. Food Chem. 1978, 26, 791.
7. Buttery, R. G.; Soderstrom, E. L.; Seifert, R. M.; Ling, L. C.; Haddon, W. F. J. Agric. Food Chem. 1980, 28, 353.
8. Stoll, M., Winter, M., Gautschi, F., Flament, I., Willhalm, B., Helv. Chim. Acta 1967, 50, 628
9. Soderstrom, E. L. Unpublished work. 1978.
10. Schreier, P. CRC Critical Reviews in Food Science and Nutrition, Nov. 1979, 59.
11. Smilanick, J. M. J. Econ. Entomol. 1979, 72, 558.

RECEIVED April 3, 1981.

5

Flavonoids, Mutagens, and Citrus

R. M. HOROWITZ

U.S. Department of Agriculture, Agricultural Research Service,
Fruit and Vegetable Chemistry Laboratory, Pasadena, CA 91106

Flavonoids--the venerable group of plant constituents so beloved by natural products chemists--are regarded by some students of the nutritional arts as beneficial, even "semi-essential" components of the diet. Whether they are truly beneficial is a question debated for decades and one that is no more likely to be solved in the foreseeable future than many other well-known and enigmatic problems in nutrition.

If flavonoids have dubious credentials as vitamins or semi-essential dietary factors, it has at least been generally agreed that, as a group, they are benign at the levels found in foods or incorporated in diet supplements and pharmaceuticals. They contain neither nitrogen nor sulfur--elements often associated with toxicity--and are thought to be metabolized chiefly to carbon dioxide and aromatic acids. Perhaps more important is the fact that since flavonoids occur in all higher plants (as well as some mosses, liverworts, fungi and ferns) they are, and always have been, a common constituent of the diet. It has been estimated that the "average" American daily diet contains about 1 gram of flavonoids (1). Even if this estimate were high by a factor of 2 to 3, the intake of flavonoids would still be substantial. To summarize, flavonoids are ubiquitous; conventional wisdom tells us they are unlikely to be harmful and may even be good for us.

A new aspect of these compounds has surfaced and it is one that will surely be visible for a long time. It has been found that certain flavonoids are mutagenic when examined in the Ames test and other short-term *in vitro* tests for mutagenicity. This finding is significant because there is substantial evidence to indicate that many known carcinogens are mutagens or are converted to mutagens by metabolic processes occurring in the body. Conversely, there is reason to believe that many mutagens are carcinogens and may also contribute to various genetic abnormalities. It is the purpose of this article to review the data on flavonoid mutagenicity and to discuss some of the implications.

The *Salmonella* test for mutagenicity

The best known and most widely used of the *in vitro* tests is that developed by Ames and co-workers (2-5). In this procedure, one of several specially constructed histidine-requiring mutants of *Salmonella typhimurium* is added to a histidine-deficient growth medium contained in a Petri dish. Since the rate of spontaneous reversion to histidine independence is low, only background growth is observed. If, in a parallel experiment, a mutagenic substance is added to the plate containing the *Salmonella* bacteria and the growth medium, mutations leading to histidine independence should occur and growth should be observed. After several days a count is made of the revertant colonies, each of which is made up of bacteria containing functional rather than defective histidine genes and each of which is descended from a back-mutated bacterium. The number of revertant colonies per nanomole of mutagen added is often taken as a measure of the mutagenic potency of the substance tested. Experience has shown that increasing the amount of mutagen in the test almost always results in a linear increase in the number of revertant colonies produced.

In practice, the experiment outlined above is not entirely satisfactory: many known carcinogens--benz[*a*]pyrene is one example--fail to show up as mutagens. The reason for the failure is that the compound in question happens not to be the proximate mutagen; it becomes a mutagen only after it enters the body and is altered by metabolic processes. In order to simulate these metabolic processes *in vitro* and thereby improve the results in the *Salmonella* test, Ames has recommended the addition of various activating agents to the test plate. Among these activating agents are homogenized liver (the S-9 fraction of rat liver microsomes induced with polychlorinated biphenyl (Aroclor) or phenobarbital) and "fecalase" (a cell-free, glycosidase-containing extract of feces). Other glycosidase-containing preparations have been made from cell-free extracts of rat cecal contents ("cecalase") and from *Aspergillus niger* ("hesperidinase"). The liver homogenates seem to carry out a variety of oxidative and hydrolytic functions; the other preparations are largely hydrolytic and their chief use is in the testing of glycosides where it is necessary to remove the sugar components in order to obtain a positive test. Clearly, in judging how well the *in vitro* *Salmonella* test mimics *in vivo* mutagenicity, much depends on the choice of activating agents. It has been pointed out that some substances may require reductive activation or may be metabolized by organs other than liver or by cell components other than microsomes (6). In some instances treatment with S-9 liver homogenate causes deactivation (7, 8, 9). Interesting discussions and rebuttals on various aspects of the subject have appeared (10, 11, 12).

Behavior of flavonoids in the *Salmonella* test

Starting in 1977, papers dealing with this topic have come from half a dozen different laboratories (13-18). The amount of data is considerable and, where the results of various investigators overlap, there is fairly substantial agreement. The kinds of flavonoids that have been studied, in free or combined form, are the flavones, isoflavones, flavonols, flavanones, dihydrochalcones, catechins and anthocyanins. Of these, only the flavonols show any appreciable mutagenicity but it must be emphasized that not all flavonols are mutagenic.

Table I contains a summary of the results reported for certain flavonols and their glycosides, all of which are or are thought to be naturally occurring. It also contains data for the pentamethyl and pentaäcetyl derivatives of quercetin, neither of which occurs naturally, and for a small group of actual or postulated metabolites of quercetin. The principal findings are these:

1) Fisetin (*1*), quercetin (*2*), rhamnetin (*3*), 5,7-di-*O*-methylquercetin (*4*) and myricetin (*5*) are all mutagenic even in the absence of any activating agent. When they are exposed to S-9 liver homogenate their mutagenic potency increases, sometimes as much as tenfold.

2) Flavonols *6-18* are not mutagenic by themselves but become so when exposed to S-9 or, in the case of glycosides, to a glycoside-hydrolyzing enzyme. (Myricitrin (*28*) would doubtlessly become active in the presence of a hydrolytic enzyme.) The remaining flavonols (*19-28*), as well as the known or postulated metabolites of quercetin (*29-33*), are inactive even if treated with S-9.

3) Quercetin (*2*) is easily the most active flavonol in the *Salmonella* test, yet it is at least an order of magnitude less potent than such familiar mutagens as benz[*a*]pyrene and chrysene. Quercetin, galangin (*6*), kaempferol (*7*) and rhamnetin (*3*) are more or less active in several other *in vitro* tests designed to detect genetic toxicity. For example, quercetin has been shown to transform hamster embryo cells (kaempferol did not) (20); to induce gene conversion in yeast and frameshift mutation in *E. coli* (17); to induce chromosome aberrations and sister chromatid exchanges in cultured human and Chinese hamster cells (21); and to give a positive response in the mouse leukemic cell mutation assay (9). In some of these tests the compound was active without addition of S-9 liver homogenate, while in others S-9 actually decreased activity.

4) Activity is restricted to but does not necessarily occur in compounds containing a free hydroxy group at position 3. Most of the active compounds have a free hydroxy group at position 5 as well, but there are exceptions, e.g., fisetin (*1*) and 5,7-di-*O*-methylquercetin (*4*), both of which have weak but definite activity. A free hydroxy group at position 7 is not essential for activity.

Table I. *The Mutagenicity of Some Naturally Occurring Flavonols and Related Substances in the Salmonella Test*[a]

Investigator →	MacGregor[b] (16)		Brown (18)			Sugimura (15)		Hardigree (17)	Bjeldanes (13)			
Tester strain TA no. →	98	100	98	100	1537	98	100	98	98	100	1535	1538
Mutagenic with or without activation												
1 Fisetin 3,7,3',4'-tetra OH	*0* 0	*0.30* 0.43	0.14			0.12	0.35	*0* 0.13				
2 Quercetin 3,5,7,3',4'-penta OH	*2.8* 12.7	*2.0* 6.9	11.6			5.3	0.35	*0.79* 2.27	*+* +++	*+* +++	*0* 0	+
3 Rhamnetin 3,5,3',4'-tetra OH-7-MeO	*0.25* 2.6	*1.6* 3.6			0.45							
4 5,7-Di-*O*-methylquercetin	*0.25* 0.63											
5 Myricetin 3,5,7,3',4',5'-hexa OH	*0.17* 1.0	*0.16* 1.1		0.12				*0.54* 1.08				
Mutagenic only if activated												
6 Galangin 3,5,7-tri OH	3.2	2.1	2.0			3.8	1.6					
7 Kaempferol 3,5,7,4'-tetra OH	2.4	1.0		7.3		3.9	0.68	4.09				

8	Robinin (kaempferol 3,7-diglycoside)			1.07^{c}							
9	Kaempferide 3,5,7-tri OH-4' MeO				0.24						
10	Quercitrin (quercetin 3-rhamnoside)	0.22		3.3^{c}			0.43				
11	Rutin (quercetin 3-rutinoside)			1.14^{c}			0.21				
12	Azaleatin 3,7,3',4'-tetra OH-5-MeO	0.11	0.16								
13	Isorhamnetin 3,5,7,4'-tetra OH-3'-MeO	0.46	0								
14	Tamarixetin 3,5,7,3'-tetra OH-4'-MeO	3.0									
15	Ombuin 3,5,3'-tri OH-7,4'-di MeO	0.60									
16	3',4'-Di-*O*-methylquercetin	0.20	0.19								
17	Quercetin pentaäcetate							+++	+++	0	0
18	Morin 3,5,7,2',4'-penta OH	0.46			0		1.21				

Continued

Table I-*Continued*

Investigator →	MacGregor[b] (16)		Brown (18)			Sugimura (15)		Hardigree (17)	Bjeldanes (13)			
Tester strain TA no. →	98	100	98	100	1537	98	100	98	98	100	1535	1538
Non-mutagenic												
19 3,7-Dihydroxyflavone	0											
20 3-*O*-Methylgalangin 5,7-di OH-3-MeO	0											
21 Resokaempferol 3,7,4'-tri OH	0	0										
22 3-*O*-Methylquercetin	0											
23 Caryatin 7,3',4'-tri OH-3,5-di MeO	0											
24 3,7-Di-*O*-methylquercetin	0											
25 3,3'-Di-*O*-methylquercetin	0											
26 Penta-*O*-methylquercetin	0								0	0	0	0
27 Robinetin 3,7,3',4',5'-penta OH	0	0	0									
28 Myricitrin (myricetin 3-rhamnoside)								0				

Quercetin metabolites or postulated metabolites									
29	3-Hydroxyphenylacetic acid					0	0	0	0
30	3,4-Dihydroxyphenylacetic acid					0	0	0	0
31	Caffeic acid	0							
32	Phloroglucinol	0	0						
33	Phloroglucinol carboxylic acid	0	0						

[a]Figures represent number of revertants per nanomole of compound tested; data in italics are from experiments in which no activating agent was used; S-9 liver homogenate was used in all other experiments except as noted; values less than 0.1 are listed as 0; TA98, TA1537 and TA1538 are frameshift tester strains; TA100 and TA1535 are base-pair substitution tester strains; flavonol numbering system is shown in structure *2*.

[b]Calculated from data in the original paper; the largest value was selected where more than one experiment was reported.

[c]Activated by cecal cell-free extract.

5) The situation with respect to substitution in the B-ring is not entirely clear. MacGregor and Jurd (16), who studied a large series of compounds and have attempted to rationalize structure-activity relations, suggest that the B-ring must be substituted in a way that allows it to become oxidized to a quinonoid intermediate. Thus, quercetin could autoxidize or be oxidized by the liver microsomal system to the quinone *34a* or the tautomeric quinone methide *34b*. Other 3',4'-dihydroxyflavonols would behave similarly. Monosubstituted derivatives could be ortho-hydroxylated by the liver preparation and methoxyl groups would presumably be demethylated at some stage. The mutagenicity of the flavonols is attributed to the fact that the derived quinone methides are alkylating agents. The idea that a quinone methide or an analogous quinone structure is the proximate mutagen receives some support from a recent finding that quinone methides such as *35* (derived from the corresponding flavylium salt) are mutagenic in the *Salmonella* test (22).

It appears that no attempts have been made to determine how liver homogenate alters flavonols, although this is a fundamental point and one that should be amenable to experiment. It is, nevertheless, indisputable that certain flavonols are mutagenic in certain *in vitro* test systems. Obviously, the question arises whether they constitute a risk in the diet of humans. Some authors are noncommittal on this important point; others have suggested that consideration be given to a breeding program aimed at eliminating or at least reducing the amount of flavonols in food plants (4, 19). This suggestion is probably unworkable, especially since we do not even know to what extent flavonols and their glycosides are essential to the economy of the plant. Moreover, a successful outcome would presumably do away with what are thought to be the beneficial effects of flavonols, i.e., protection of lipids and ascorbic acid from oxidation. In the sections below we discuss some factors that may be useful in evaluating the possible risk of ingesting these compounds.

Flavonol occurrence and metabolism

Extrapolation of the results obtained from surveys of leaf and flower petal constituents has led to the conclusion that practically all higher plants are able to synthesize flavonols and that plant tissues totally lacking in these substances are rare (23). Reported absences may in most cases simply indicate that insufficiently sensitive methods were used. Quercetin and kaempferol are by far the most common flavonols and they are almost always found in the form of glycosides. In general, leaves, flowers, fruit and other living tissues contain glycosides; woody tissues tend to contain aglycones and seeds may contain either. Because light is required in the biosynthetic process, flavonol glycosides are usually found in highest concentra-

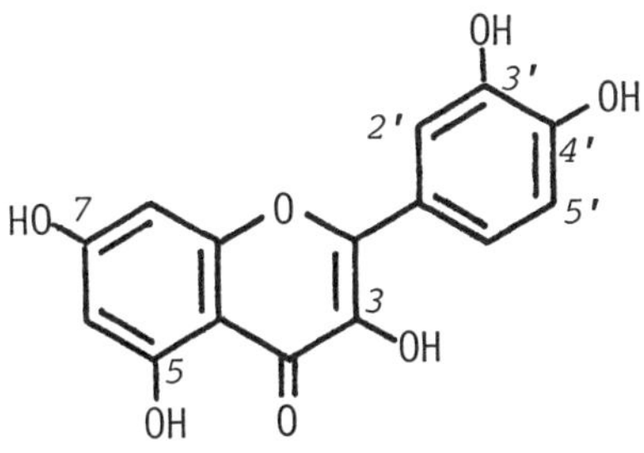

2 Quercetin

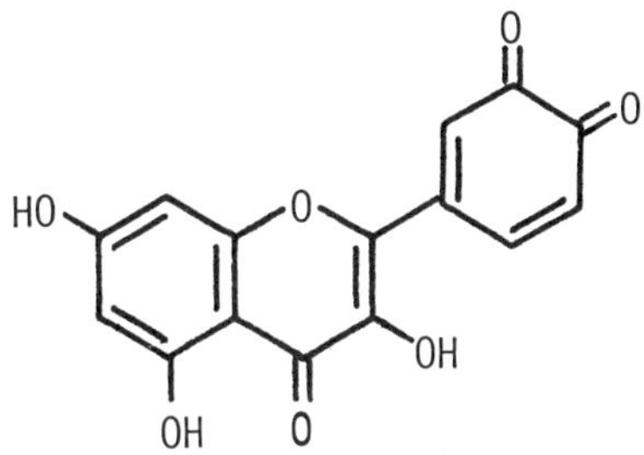

34a

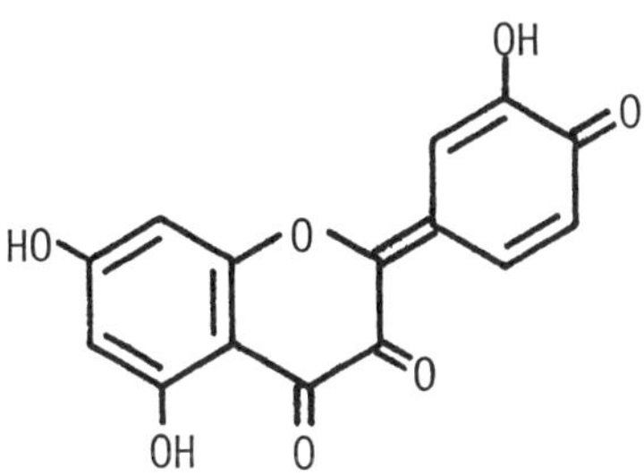

34b

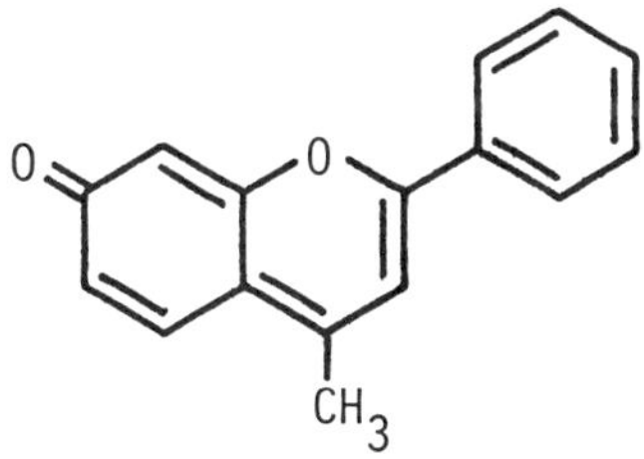

35

tion in the skins and outer leaves of fruits and vegetables and in lowest concentration in root vegetables and other underground parts of plants. In a recent review (24), glycosides of quercetin or kaempferol were reported to occur in every one of a group of about 40 different food plants comprising fruits, berries and vegetables of all types. Thus, it appears that the ingestion of some quantity of quercetin and kaempferol glycosides is unavoidable, while the ingestion of any quantity of quercetin and kaempferol aglycones would be unusual.

Since aglycones but not glycosides are mutagenic, we may well wonder whether glycosides are hydrolyzed to aglycones after ingestion. The usual answer to this question is that hydrolysis does occur and the usual evidence is that cecal or fecal microflora cause splitting of the sugars of plant glycosides when tested *in vitro*. It is, however, by no means certain that flavonoids invariably undergo hydrolysis to the aglycone or, if they do, that the aglycone persists for an appreciable length of time. In a paper dealing with the effects of isolated rat cecal microflora on flavonoid glycosides and other compounds, Scheline (25) pointed out that only some of the reactions occurring in the gut are revealed by *in vitro* experiments, since perhaps the majority of microorganisms growing in the gut require special conditions for their culture. Moreover, the fact that a reaction occurs in an incubate may not accurately reflect its quantitative significance in the gut, because the growth of minor organisms may be favored under the artificial conditions of the experiment. *In vitro* studies on a rumen microorganism, *Butyrivibrio* sp. C_3, have shown (26) that it anaerobically degrades rutin (*11*) or quercitrin (*10*), but not the free aglycone. At low temperature (4°) quercetin accumulates in the medium while at higher temperature (39°) the glycosides disappear and the aglycone fails to accumulate. (The products are phloroglucinol (*32*), carbon dioxide, 3,4-dihydroxybenzaldehyde and 3,4-dihydroxyphenylacetic acid (*30*)). When quercetin is provided as the substrate, very little is used. It is thought that the *Butyrivibrio* glycosidases are intracellular and the block in the use of free aglycone is a result of its insolubility and inability to enter the cell. In an earlier study (27) it was found that the molds *Aspergillus flavus* and *A. niger* produce extracellular enzymes capable of degrading rutin; the products are rutinose, protocatechuic acid, phloroglucinol carboxylic acid (*33*) and a phloroglucinol carboxylic acid-protocatechuic acid ester. Since quercetin is not detected as an intermediate in the degradation, it was suggested that one possible mode of action might be cleavage of the heterocyclic ring of the intact glycoside rather than hydrolysis of the sugars.

Quercetin (except possibly in traces) has not been detected as a metabolite of rutin after feeding it to rats, rabbits, guinea pigs and humans. The only metabolic products reported are 3,4-dihydroxyphenylacetic acid (*30*), 3-methoxy-4-hydroxyphenylacetic acid, 3-hydroxyphenylacetic acid (*29*) and methylcatechol glucuronide (28, 29, 30). The same metabolites are found when quercetin is fed instead of rutin. In experiments with small (5 mg) oral doses of ^{14}C-labeled quercetin none of the unchanged compound could be found either in the urine or gastrointestinal contents of rats (31). The glycoside myricitrin (*28*) yields 3,5-dihydroxyphenylacetic acid as the major product in both urine and feces at a dose of 100 mg/rat; some myricetin (5) is found in the urine but none in the feces (32). A study of the ^{14}C-labeled flavanone glycoside, neohesperidin, and its reduction product, the sweetener neohesperidin dihydrochalcone, has shown that, although metabolism takes place, neither of the corresponding aglycones, hesperetin and hesperetin dihydrochalcone, occurs in free or combined form in the urine of rats that have been fed these substances (33). In germ-free rats flavonoids fail to undergo the ring-fission reactions that yield phenylacetic acids and related compounds (30, 32).

It seems reasonable to conclude that the metabolism of quercetin glycosides is brought about largely by the action of intestinal microflora and that more than one pathway is involved. One of these may be initial hydrolysis of the glycoside to the aglycone and another may be a direct attack on the flavonol moiety of the intact glycoside. In any event, the available data suggest that the aglycone has only a fleeting existence before it is further degraded to aromatic acids and carbon dioxide.

Feeding studies

Several chronic toxicity studies on rutin, quercitrin and quercetin have been published. A brief summary of the experiments is given in Table II.

In the early work of Wilson (34) and Ambrose (35) no evidence was obtained that indicated any appreciable toxicity of rutin, quercitrin or quercetin at a 1% dose level in rats. Histopathology was carried out in these studies but no detailed summary of the results was published. In a recent paper Saito and co-workers (37) reported a longer (∿2-year) study of quercetin in mice at a 2% dose level. There was a fairly high incidence of tumors in both treated and control animals and some of the tumors in the treated group were of unusual types. (The quercetin in this experiment was obtained commercially and used without further purification.) It was concluded that there is no indication that quercetin is a potent carcinogen. In contrast, Pamucku and co-workers (36) recently published the results of a 1-year rat study in which a high incidence of intestinal and bladder tumors was found in animals that were kept on a grain diet containing 0.1% quercetin. (It was claimed that the quercetin was more than 99% pure.) The results are all the more astonishing because the dose was only about 1/10 that used in the Ambrose experiment and 1/20 that used in the Saito experiment.

Clearly, it is difficult to reconcile the discrepancies in

Table II. *Chronic Toxicity Experiments on Rutin, Quercitrin and Quercetin*

Investigator	Animals; maximum number per group	Diet	Compound; highest dose	Duration in days	Total amount ingested per animal[a]	Results
Wilson (1947) (34)	Rats; 6	Purina Dog Chow	Rutin 1%	400	70 g	Growth, organ weights, histopathology and reproduction normal
Ambrose (1952) (35)	Rats; 5	See b	Quercitrin 1%	410	66 g	Growth, organ weights, blood chemistry and histopathology normal
Ambrose (1952) (35)	Rats; 5	See b	Quercetin 1%	410	66 g	Growth, organ weights, blood chemistry and histopathology normal
Pamukcu (1980) (36)	Rats; 18	Grain	Quercetin 0.1%	406	8 g	Controls: no tumors Treated: 80% developed intestinal tumors; 20% developed bladder tumors
Saito (1980) (37)	Mice; 35-38	Basal pellet diet CE-2	Quercetin 2%	842		No significant difference in incidence of tumors in control and treated groups; several unusual tumors in treated group

[a]Rough estimates calculated from data given in the original papers.

[b]Casein 10%, corn meal 73%, linseed oil meal 10%, alfalfa 2%, bone ash 1.5%, sodium chloride 0.5%, cod liver oil 3%.

these studies but factors such as purity, solubility and moisture give rise to worrisome questions. If a grain ration were inadvertently exposed to moisture could mycotoxins be formed? What is the effect of the presumably insoluble, crystalline quercetin used in all these experiments? Microbial action *in vitro* is extremely slow on crystalline quercetin (26). In humans given 4 grams of quercetin orally more than half of the dose is excreted unchanged and there is no evidence of appreciable absorption from the gut (38).

The feeding studies do not really tell us whether quercetin is a carcinogen, although the balance of evidence suggests it is not. It seems clear that much of the effort has gone into the wrong compound. Rutin or some other naturally occurring glycoside of quercetin would be more likely than quercetin itself to give a reliable answer to the question of possible carcinogenic flavonols in foods of plant origin.

Flavonoids in *Citrus*

Citrus fruit is a major source of flavonoids in the human diet; the peel is a major source of the flavonoids that find their way into diet supplements and pharmaceuticals. (It may be noted, incidentally, that either wet or dried peel is used as cattle feed.) Flavanones, flavones, flavonols, dihydrochalcones and anthocyanins are the main types of flavonoids in *Citrus* (39). They occur in highest concentration in the peel and to a lesser extent in the juice and edible part of the fruit. By far the most abundant flavonoids in *Citrus* are flavanone glycosides, e.g., hesperidin in oranges and lemons, eriocitrin in lemons, and naringin and poncirin in grapefruit. Hesperidin and naringin are easily isolated and are sold commercially, the latter for use as a bittering agent. Both compounds can serve as starting materials for the manufacture of various dihydrochalcone sweeteners. Flavones are numerous and some of them, such as diosmin in lemons or rhoifolin in grapefruit, are present in substantial quantity. They occur either as glycosides, *C*-glycosyl derivatives or permethyl ether derivatives. Dihydrochalcones have thus far been found only in kumquats and only in the form of *C*-glycosyl derivatives. Anthocyanins occur chiefly as constituents of the blood oranges to which they contribute the characteristic color.

Flavonols, the group of flavonoids most likely to contain mutagens, are represented in *Citrus* by a small number of examples. These include rutin (*11*), possibly other glycosides of quercetin (*2*), a glycoside of kaempferol (*7*) and a glycoside of isorhamnetin (*13*). With the exception of rutin, which has been obtained in 3% yield from the green fruit of the Satsumelo (a hybrid of grapefruit and tangerine), all of these presumably promutagenic compounds are relatively minor constituents. Few quantitative studies have been published, but it has been estimated that, after enzymatic hydrolysis of the natural glycosides, 100 ml of single-strength lemon juice contains about 2.2 mg of quercetin in addition to 20 mg of eriodictyol and 1.4 mg of hesperetin, both of which are flavanones (40). Other flavonol glycosides in *Citrus* are the 3-β-D-glucosides of limocitrin, limocitrol and isolimocitrol. Nothing is known about the possible mutagenicity of these minor glycosides or their aglycones.

Citrus fruits invariably contain complex arrays of flavonoid compounds, the most important of which, both in number and quantity, are flavanones and flavones. A total of about 35 flavanones and flavones have been identified, compared to about 6 flavonols (39). Since flavanones and flavones are not mutagens, the mutagenic potential of crude, hydrolyzed preparations of citrus flavonoids is likely to be small. Moreover, there are indications that the mutagenicity of flavonols may be modified by the presence of other flavonoids. Thus, Brown and Dietrich (18) showed that the activity of galangin (*6*) in the *Salmonella* test is strongly suppressed by the addition of either the flavone apigenin or the flavonol robinetin (*27*), although neither of these compounds was more than minimally effective in reducing quercetin mutagenicity.

Perhaps the most thoroughly tested of all flavonoids is the sweetener neohesperidin dihydrochalcone, which is formed by reducing the flavanone glycoside neohesperidin, a constituent of Seville oranges. The dihydrochalcone, after 2-year feeding tests at various dose levels in rats and dogs, gave no evidence of tumor induction or teratogenicity (41). The compound has also been checked (a) in mice to see whether it causes an increase in the normal frequency of micronucleated polychromatic erythrocytes in bone marrow (42) and (b) in *Salmonella* tester strains TA98, TA100, TA1535, TA1536, TA1537 and TA1538 to see whether it causes reversion to histidine independence (14, 16, 18, 42, 43). Again, there was no evidence of mutagenicity.

Are flavonols a menace?

Red wine (but not white), black tea and green tea (but not coffee), pickles and grapejuice--all of these, we are warned, are mutagenic in the *Salmonella* test (5, 44). Some of them are active even without hydrolysis. We are told that "glycosides of quercetin, a mutagenic flavonoid, are present in considerable amounts in our diet from a variety of sources.....and, by means of hydrolysis, bacteria in the human gut readily liberate the mutagen" (3). The implication seems clear enough: foods and beverages of plant origin may conceivably be a hazard to health. Before accepting this view let us summarize what has been discussed above.

The natural state of flavonols in foods is almost exclusively a nonmutagenic glycoside. Hydrolysis liberates the free flavonols, some of which are mutagenic. Although certain gut microflora are able to cleave glycosides to aglycones, other pathways may also be operative whereby the aglycone moiety of the intact glycoside is degraded directly without the free aglycone ever being liberated. In any event, studies of the metabolism of quercetin glycoside have, thus far, provided no evidence for the accumulation of free quercetin; instead, non-mutagenic aromatic acids resulting from the degradation of quercetin are found. When free quercetin is fed to humans no detectable quantity is absorbed through the gut; most of the quercetin passes through unchanged and the rest presumably undergoes bacterial degradation. Animal feeding studies on quercetin glycosides have given no evidence of carcinogenicity; feeding studies of quercetin itself have given conflicting results. Finally, on a more speculative level, Cairns (45) has noted that "presumably there has always been strong selection pressure for adequate methods of DNA repair, and so it is hardly surprising to find such low levels of risk from agents like UV light and X-rays, which have not changed in intensity for millions of years." Since humans have always been exposed to flavonol precursors, is it not reasonable to assume that here too we are not likely to exceed our DNA repair capabilities?

More information on these complex subjects will doubtlessly be forthcoming but, on the basis of what we know now, it is difficult to reach any conclusion other than that the risks associated with the ingestion of flavonols are minimal.

Addendum

After this article was submitted for publication the results of a new feeding study were reported by Hirono and co-workers (46). In one experiment ACI rats were kept for 540 days on basal pellet diet CE-2 containing either 1% quercetin, 5% quercetin or 5% rutin; in a second experiment ACI rats were given either 10% quercetin or 10% rutin for 850 days. A total of 256 animals were used (including controls). Most tumors found in the experimental groups were also present in the corresponding control groups and there was no significant difference between the incidence of tumors in the control and experimental groups. The authors concluded that quercetin and rutin are not carcinogenic in ACI rats.

LITERATURE CITED

1. Kühnau, J. World Review of Nutrition and Dietetics 1976, 24, 117-191.
2. Ames, B. N.; McCann, J.; Yamasaki, E. Mutation Research 1975, 31, 347-364.
3. Ames, B. N. Science 1979, 204, 587-593.
4. Ames, Bruce in "Genes, Cells and Behavior" edited by N. H. Horowitz and E. Hutchings, Jr.; W. H. Freeman and Company: San Francisco, 1980; pp 21-31.

5. Tamura, G.; Gold, C.; Ferro-Luzzi, A.; Ames, B. N. Proceedings of the National Academy of Sciences 1980, 77, 4961-4965.
6. Bridges, B. A. Nature 1976, 261, 195-200.
7. DeFlora, S. Nature 1978, 271, 455-456.
8. Aeschbacher, H. U.; Chappins, C.; Wuerzner, H. P. Food and Cosmetics Toxicology 1980, 18, 605-613.
9. Meltz, M. L.; MacGregor, J. T. Mutation Research 1981, 88, 317-324.
10. Ashby, J.; Styles, J. A. Nature 1978, 261, 452-455.
11. Ames, B. N.; Hooper, K. Nature 1978, 274, 19-20.
12. Ashby, J.; Styles, J. A. Nature 1978, 274, 20-22.
13. Bjeldanes, L. F.; Chang, G. W. Science 1977, 197, 577-578.
14. Brown, J. P.; Dietrich, P. S.; Brown, R. J. Biochemical Society Transactions 1977, 5, 1489-1492.
15. Sugimura, T.; Nagao, M.; Matsushima, T.; Yahagi, T.; Seino, Y.; Shirai, A.; Sawamura, M.; Natori, S.; Yoshihira, K.; Fukuoka, M.; Kuroyanagi, M. Proceedings of the Japan Academy, Ser. B. 1977, 53, 194-197.
16. MacGregor, J. T.; Jurd, L. Mutation Research 1978, 54, 297-309.
17. Hardigree, A. A.; Epler, J. L. Mutation Research 1978, 58, 231-239.
18. Brown, J. P.; Dietrich, P. S. Mutation Research 1979, 66, 223-240.
19. Brown, J. P. Mutation Research 1980, 75, 243-277.
20. Umezawa, K.; Matsushima, T.; Sugimura, T.; Hirakawa, T; Tanaka, M.; Katoh, Y.; Takayama, S. Toxicology Letters 1977, 1, 175-178.
21. Yoshida, M. A.; Sasaki, M.; Sugimura, K.; Kawachi, T. Proceedings of the Japan Academy, Ser. B. 1980, 56, 443-447.
22. Sweeny, J. G.; Iacobucci, G. A.; Brusick, D.; Jagannath, D. R. 181st Meeting of the American Chemical Society, Atlanta, Georgia, March 29-April 3, 1981; Abstract AGFD 24.
23. Harborne, J. B. "Comparative Biochemistry of the Flavonoids"; Academic Press: London and New York, 1967; p 110.
24. Herrmann, K. Journal of Food Technology 1976, 11, 433-448.
25. Scheline, R. R. Acta Pharmacologica et Toxicologica 1968, 26, 332-342.
26. Krishnamurty, H. G.; Cheng, K.-J.; Jones, G. A.; Simpson, F. J.; Watkin, J. E. Canadian Journal of Microbiology 1970, 16, 759-767.
27. Westlake, D. W. S.; Talbot, G.; Blakley, E. R.; Simpson, F. J. Canadian Journal of Microbiology 1959, 5, 621-629.
28. Murray, C. W.; Booth, A. N.; DeEds, F.; Jones, F. T. Journal of the American Pharmaceutical Association, Scientific Edition 1954, 43, 361-364.
29. Booth, A. N.; Murray, C. W.; Jones, F. T.; DeEds, F. Journal of Biological Chemistry 1956, 223, 251-257.

30. Griffiths, L. A. in "Topics in Flavonoid Chemistry and Biochemistry" edited by L. Farkas, M. Gábor and F. Kállay; Elsevier Scientific Publishing Company: Amsterdam, 1975; pp 201-213.
31. Petrakis, P. L.; Kallianos, A. G.; Wender, S. H.; Shetlar, M. R. Archives of Biochemistry and Biophysics 1959, 85, 264-271.
32. Griffiths, L. A.; Smith, G. E. Biochemical Journal 1972, 130, 141-151.
33. Horowitz, R. M.; Gentili, B. Citrus Research Conference, U.S. Department of Agriculture, Pasadena, California, December 8, 1976; Abstracts of Papers, pp 8-9.
34. Wilson, R. H.; Mortarotti, T. G.; Doxtader, E. K. Proceedings of the Society for Experimental Biology and Medicine 1947, 64, 324-327.
35. Ambrose, A. M.; Robbins, D. J.; DeEds, F. Journal of the American Pharmaceutical Association, Scientific Edition 1952, 41, 119-122.
36. Pamukcu, A. M.; Yalciner, S.; Hatcher, J. F.; Bryan, G. T. Cancer Research 1980, 40, 3468-3472.
37. Saito, D.; Shirai, A.; Matsushima, T.; Sugimura, T.; Hirono, I. Teratogenesis, Carcinogenesis and Mutagenesis 1980, 1, 213-221.
38. Gugler, R.; Leschik, M.; Dengler, H. J. European Journal of Clinical Pharmacology 1975, 9, 229-234.
39. Horowitz, R. M.; Gentili, B. in "Citrus Science and Technology," volume 1 edited by S. Nagy, P. E. Shaw and M. K. Veldhuis; Avi Publishing Company: Westport, Connecticut, 1977; pp 397-426.
40. Vandercook, C. E.; Stephenson, R. G. Journal of Agricultural and Food Chemistry 1966, 14, 450-454.
41. Gumbmann, M. R.; Gould, D. H.; Robbins, D. J.; Booth, A. N. in "Sweeteners and Dental Caries" edited by J. H. Shaw and G. G. Roussos; Information Retrieval, Inc.: Washington, D.C., 1978; pp 301-310.
42. MacGregor, J. T. Toxicology and Applied Pharmacology 1979, 48, A47.
43. Batzinger, R. P.; Ou, S.-Y. L.; Bueding, E. Science 1977, 198, 944-946.
44. Uyeta, M.; Taue, S.; Mazaki, M. Mutation Research 1981, 88, 233-240.
45. Cairns, J. Nature 1981, 289, 353-357.
46. Hirono, I.; Ueno, I.; Hosaka, S.; Takanishi, H.; Matsushima, T.; Sugimura, T.; Natori, S. Cancer Letters 1981, 13, 15-21.

RECEIVED June 26, 1981.

6

Apple Quality Characteristics as Related to Various Processed Products

R. L. LA BELLE

New York State Agricultural Experiment Station, Cornell University, Geneva, NY 14456

Apples, by virtue of being widely grown, amenable to long storage, and both bland of flavor and firm of texture, are possibly processed into a wider array of products than any other fruit. While these might be roughly classified as liquid, solid, and pureed (Table 1), a diversity of product type exists within each of these categories. In addition, there are differences even within products, arising from the particular processing method used, that also affect the characteristics desired in the raw product.

This situation has led to some confusion in evaluating apple quality for processing and to uncertainty in establishing its market value. Standards, whether of federal or state origin, have been too general to serve very well the individual users concerned with a particular product or process. Consequently, there have been attempts to set up more explicit grades that better fit

Table I. Classification of Products Processed from Apples.

LIQUID	SOLID	PUREED
cider (fresh)	baked whole apple	applesauce (including baby food)
juice (fresh or pasteurized)	slices or pieces:	nectar
juice concentrate (4:1 or 7:1)	fresh	apple butter
hard (fermented) cider	frozen	
wine	canned	
vinegar	dried	
jelly	dehydrofrozen	
essence	rings (canned, spiced)	

0097-6156/81/0170-0061$05.00/0

existing needs (1), sometimes even on an individual processing plant basis. On the other hand, the potential for utilizing apples of certain quality characteristics for the most appropriate of these diverse products has been at times a distinct marketing advantage. This paper will review the raw product quality requirements of the major processed apple products and consider the pertinent measurements that may be made on fresh apples.

Apple Characteristics

It will be useful first to consider the many characteristics in which apples may differ on account of variety, maturity, or post-harvest condition. Most people are familiar with and able to identify by sight several of the varieties commonly marketed or grown in their own region, and to remember a few of their more obvious quality attributes as well. They might even have some idea of how these factors change as the fruit progresses through maturation on the tree or ripens after harvest. The International Apple Institute lists only 14 major varieties in a recent survey of storage holdings nationwide (2), and these would be the best known.

Growers and processors of apples are especially knowledgeable about the characteristics of importance to final product yield and quality among these varieties. However, in practice only a few of these factors are measured and taken into account in assigning market value or in directing the fruit to usage in one product or another. In research we have been more painstaking, no doubt to the point of impracticality, in defining and measuring apple quality for processing, as discussed at length in this paper. Regardless of whether all such factors are actually measured, an awareness of their effects can be valuable to the industry.

Damage and decay. Damage to the fruit while still on the tree may result from insects, disease, weather (including hail and wind), and machine operations - which together result in misshapen fruit, scar tissue, and possibly insect residue. We might even include external spray residue in this category. Some of the damaged fruit, especially if decayed, will fall to the ground and be fortuitously excluded from the harvest (unless picked up again by man or machine). The relatively rough harvest by hand of fruit destined for processing ordinarily results in many shallow bruises, while machine-harvest by shaking the tree usually inflicts a greater number of cuts and punctures. The bruises don't worsen during storage of the harvested fruit, but skin-breaks permit microbial entry that, depending very much on storage temperature, leads in time to rot (decay, or spoilage). The bruised, discolored tissue and minor decay may be trimmed or screened out during processing, affecting yield only slightly though increasing labor-cost; but extensive decay may interfere with peeling, leading to total loss of the fruit, or affect the flavor, color, or shelf-life of a juice or puree, if not removed. Losses of "utilizable material" may be as high as 20 percent in badly handled and indifferently stored apples (3).

Size and shape. Apple size has been traditionally expressed as diameter ranges (in fractional inches) based on the openings in the endless-conveyor grading chains used in the plant. Pricing of these successive size-grades is commonly stepped down at the small end. The predominant size is 2-3/4 to 3 inches, and a pricing penalty may be established below this or some smaller size - typically a 2-1/2-inch minimum. Alternatively, the number of fruit in a hundredweight may be counted, reducing size identification to a single number ranging, among eleven commercially-important varieties in New York (4), from about 200 to 400 (apples per cwt). In research size may also be expressed as the unit weight in grams of a single fruit (from 100 to 200 g).

Apple shape varies from oblate to conic, expressed most simply by an increase in ratio of height to diameter in the range .80 to .95. In the high range there may be difficulty in orienting the fruit in automatic peeling machinery. Deviations from the desirable uniform roundness, such as pronounced ribbing and eccentric or skewed core axis, are more difficult to evaluate but probably not common or important enough to bother with. The diameter of the five-lobed array of seed pockets, lined with hard, unpalatable carpel tissue, does vary among cultivars to an important extent (18 to 27 mm) and is a factor in the completeness of its removal from the product.

Specific gravity. Apples differ from most fruit in having a random network of voids or passages among the tissue cells, amounting to as much as 20 to 25 per cent of tissue volume (5). Together with the seed pocket, this space, filled with air or respiratory gasses, lowers the specific gravity well below 1.0 (typically .75 to .85) and accounts for the familiar observation that apples bob about in water, as do also slices prepared from them. This attribute has several important ramifications, to be discussed. Specific gravity is strongly influenced by apple variety, most familiarly affecting weight per bushel, although this is not in itself of profound processing significance.

Color. Skin-color plays only a minor role, though limiting the usefulness of some red varieties for processing. Red skin imparts color to the flesh when cooked with it and even, if a particularly "bleeding red" that penetrates threadlike into the flesh, to the usual cold-pressed juice. Since the red anthocyanin pigments are unstable, turning brown during the usual two-year shelf-life of unrefrigerated canned products, it's not even a practical way to obtain an intentionally pink sauce or juice. Categorization of apples as red-skinned, as opposed to the more inocuous green or yellow, is usually sufficient description, with little necessity for more closely quantifying the redness.

Flesh-color varies from nearly white, sometimes tinged with green, through various intensities of yellow. This may readily be measured as the hue, chroma, and brightness of reflected color, as with the Hunter Color Meter (Hunterlab, 9529 Lee Highway, Fairfax, VA 22030), if precautions are taken against the sometimes rapid darkening of freshly-cut apple tissue. These three parameters of flesh-color have a direct bearing on the resulting product color, guidelines for which may vary from one processor or marketing area to another. A light, bright yellow is usually but not universally preferred. Ranges of the Hunter color values are: hue (a/b) = (-).05-.45; chroma ($\sqrt{a^2 + b^2}$) = 15-32; brightness (L) = 82-91.

Firmness (pressure test). The texture of fresh apples has been measured for fifty years by use of the simple spring-loaded penetrometer developed by Magness and Taylor in 1925 (6). Two commercial versions are in common use both in research and industry: the linear-scale instrument of the Ballauf Mfg. Co. and the circular-dial "Effegi." of Italian manufacture. While the "pressure-tester" has frequently been used as a guide to readiness for harvest (maturity), it is much more dependable in monitoring post-harvest ripening, or in defining the condition of the fruit after harvest as a guide to sale or processing use. This follows from the typical shape of the curve of pressure-test reading (in lb or kg) against time - nearly flat with only a slight negative slope while decreasing to about 20 lb/9 kg through maturation and harvest, then steeply negative during the middle stage of ripening and leveling off again to approach some low level of firmness (9 lb/ 4 kg) assymptotically. In commerce the values are still commonly given in lb.

"Pressure test" is something of a misnomer, as it is a force, equal to the resistance to shear and crushing under the slightly convex 7/16-inch diameter tip, that is measured and not pressure in the usual physical or engineering sense. Its widespread use can be explained by its simplicity and easy adaptability to use in the orchard or plant. Furthermore, it does provide a reasonably good correlation - somewhat dependent on variety - with both peelability and final product texture in the case of apple slices or sauce and with pressability for cider and juice. In industry a common mistake is to include too few apples in the sample evaluated, 10 to 25 being a useful sample size (7). The usual processing range is 11 to 18 lb/5 to 8 kg.

Soluble and total solids. The common use of soluble solids to define fresh apple quality came about in response to the need for a more reliable and meaningful guide to readiness for harvest. A useful estimate of percent soluble solids - mostly sugars, but also including the usual 1/4 to one percent of organic acid - is readily determined as the equivalent °Brix, even in the field, by a representative drop of juice in a hand refractometer. The usual range is 10 to 15° Brix. Total solids, or dry weight, is measured by drying a very small aliquot of dilute pureeed material in a vacuum oven, although useful estimates with perhaps less sampling error have been obtained in our laboratory by freeze-drying larger samples (300 g) to about .5% moisture. Total solids average 1.4% greater than soluble solids, varying from this in a pattern not yet well-defined.

Acidity and pH. These chemical and flavor factors are easily determined in the laboratory, if not in the field, and are frequently used in the processing plant as well as in research to evaluate both apple condition and potential product flavor. The total titratable acidity (principally malic), starting at a level characteristic of the variety, progressively drops off during both maturation and ripening to provide a useful index of those successive developments. Since acidity is changing in the opposite direction from soluble solids content during maturation, the ratio of the two, variously referred to as "sugar-acid", "Brix-acid", or "soluble solids-acid", shifts rapidly and is more useful as a guide to readiness for harvest than is pressure test (8). However, the familiar caveat concerning seasonal variation must be given; the levels of either or both parameters of the ratio may differ appreciably from one year to the next, even in fruit from a particular orchard, so that bench-marks should be established early in the season for the ensuing changes during maturation.

The measurement of pH seems more readily and accurately accomplished with present meters than a generation ago, and it is commonly used to monitor the endpoint of the acid-base titration in place of the color change of an indicator. Its relative role in the perception of acidity or tartness has been recently studied in wine from grapes (9), where the mix of organic acids (malic, tartaric, and citric) is different, but this needs further definition in apple. In wine it is secondary to total acidity in its effect on flavor. The pH value usually lies between 3.2 and 3.8, with some excursions, even among apple varieties used in commerce, outside that range. On the high side such abberations may signal that the usual means of preservation of acid foods - for example, hot-fill without subsequent retorting - may not suffice.

Acidity is of special importance to the flavor of processed apple products in that, like sweetness, it remains substantially unchanged during normal canning, freezing, or drying, while more subtle volatile flavor elements are likely to degrade or disappear.

Other flavor elements. Fresh apple flavor is characterized by what is apparently a large number of organic compounds, present in trace amounts, that are most readily determined by chromatographic methods (10). Many of these volatile compounds will be greatly reduced or absent in the processed product, unless replaced by the essence obtained by stripping in the initial stages of juice concentration. Varieties having familiar volatile flavor notes that distinguish them when fresh are thereby reduced to a more bland apple flavor, together with a characteristic balance of sweetness and tartness. However, this has its advantages in the market place, where uniformity rather than diversity is ordinarily desired, offending no one if falling somewhat short of premium quality. Laboratory evaluation of apple flavor is still a relatively undeveloped science, but there is little incentive to identify and protect these trace flavor elements as matters now stand. Probably no concensus exists for what constitutes high quality in apple flavor despite comments, not uncommonly heard, deploring its absence.

Total tannin or polyphenols can be determined in apple by permanganate titration without too much difficulty, and this has some usefulness in evaluating the level of astringency that plays a traditional role in fermented cider, particularly of the British or European type. However, the chemistry seems not yet to be well-defined (11, 12), so that the various roles of the tannic materials - providing the sensation of astringency and sometimes a bitter flavor note; causing more or less browning of color; assisting in auto-clarification of juice, cider, and wine; and either promoting or preventing undesirable hazes - are not well predicted by the usual tests. The importance of this chemistry seems likely to grow along with the sophistication of processing and consumer products. Meanwhile, the less said here by someone who is not a chemist, the better for the reader.

Rate of browning. Oxidation of polyphenol substrate in the presence of active polyphenolase and oxygen causes browning of freshly-cut or crushed apple tissue that may be discernible in less than five minutes at room temperature. There is considerable variation among, and even within, cultivars in the rate at which browning proceeds due to differences in the level or chemical nature of the substrate or of the enzyme itself; it has not been made clear which is the more likely constraint. The rate of browning on a freshly-cut surface of apple flesh has been determined in terms of the change in reflectance (ΔL) with the Hunter color meter and found to range from 1 to 8 units in 10 min. Oxidative browning may be no drawback to such products as cider, apple butter, or vinegar but is undesirable in slices, sauce, juice concentrate, and some styles of juice. Fortunately, it can be controlled by rapid thermal inactivation of the enzymes, by chemical inhibition - as with ascorbic acid or sulfur dioxide, or to a lesser extent by excluding oxygen, maintaining low temperature, or minimizing the extent of ruptured cells. Apple juice color is affected both by varietal flesh color and by the degree of browning permitted during processing. The intensity of stabilized brown color in a clear liquid product can be measured spectrophotometrically as absorbance of transmitted light and expressed as optical density @ 410 nm or merely compared to grading standards, such as the USDA Honey Color Comparometer.

Juiciness. The ready separation of juice from the flesh, touted as desirable in an apple eaten out of hand, might also be advantageous in the pressing of apples for juice, but not to the slice- or sauce-packer. While some indirect attention has been paid to it in observations of pressability, sauce consistency and yield, and the drained weight of fresh or frozen slices, apparently no one has bothered to measure juice separation from freshly cut or crushed apple tissue directly in a laboratory procedure. Varieties are generally known to lean one way or the other in the tendency to be juicy and to be valued accordingly by the processor for the effect it might have on yield or quality. Also, ripening has up to a point, the effect of increasing juice separation. It's possible that the development of an appropriate laboratory test would lead to interesting and more accurately known correlations.

Quality Factors Important to Different Processed Products

What the raw product should be like to provide suitable yield and quality is dictated as much as anything by the final form of the processed product. The unit operations themselves and the nature of the machines and processing media or chemicals used to implement them have additional requirements and effects. Together they determine, for instance, whether there will be opportunity to exclude undesirable or damaged parts of the fruit, whether the texture needs to be extra firm or the solids content preferably high, and whether such factors as browning rate or juiciness are likely to be problems. Each product seems to have a different combination of requirements, as shown by the ratings in Table 2 and amplified in the text to follow.

Baked whole apple. Although a specialty item of low volume and small importance, it's included in this presentation as the sole representative of the nearly whole apple (if we may ignore the candied product). Uniformity of size and firm, cohesive texture seem to be paramount considerations. Along with the core, about two-thirds of the peel should be removed, starting from the smaller blossom-end. This aids in holding the apple together when cooked but avoids the problems with skin that is either too tough to cut with a spoon or that splits at random. Flesh and skin color are probably matters of taste, and the degree of darkening is relatively unimportant, unless the prepared apples are frozen for baking later. A small seed pocket and a non-juicy character would be advantageous. Appreciable tartness is also desirable to offset the considerable amount of sweetener added in preparation for the table.

Table 2. Importance of Various Quality Factors to Different Processed Products.

Quality factor	Desired level	Baked apple	Fresh or Frozen	Canned	Dried or Dehydro-frozen	Apple sauce	Cider & Juice	Juice oncen-trate	Hard cider & Wine
Variety	[a]	++	++	++	++	+			+
Maturity	full		++			+	++	+	++
Condition:									
ripeness	medium			-			+		++
damage	low	+++	+++	++	++	+			
decay	low	+++	+++	++	++	++	++	++	+
Size	med.-lg.	+	++	++	++[b]	++			
Shape (ht./diam.)	low		++	++	+	+			
Seed pocket size	small	+++	+++	++	++				
Specific gravity	high		-	++	-				
Skin color	not red		+		+	++			
Flesh color	yellow		++	+	+	+++	+		

Table 2 (cont.)

Quality factor	Desired level	Baked apple	Fresh or Frozen	Canned	Dried or Dehydro-frozen	Apple sauce	Cider & Juice	Juice oncen-trate	Hard cider & Wine
Firmness (pressure test)	firm	+++	++	+++	++	+			
Soluble solids	high	+			+++	++	+	+++	
Total solids	high				+++	++			
Total acid	medium	++	+	++		++		-	-
pH	medium	++	+	++		++		-	-
Volatile flavor	medium	++	+			++	++	+	
Tannins	low								--
Rate of browning	low		++	+	+++	++		+	
Juiciness	medium	--	-		-	-	+	+	+

\+ ++ +++ = degree of importance
\- -- = importance opposite to normally desired level indicated in the second column

a) varietal importance is based on the usual practice of marketing as single varieties without blending
b) if diced rather than sliced, the larger the better, so long as still machine-peelable

Fresh or frozen slices. Fully mature, well-ripened apples of a single variety are required to provide the premium quality expected of the frozen product. To a lesser extent this is also true of the fresh slices that are lightly treated with antioxidant for a shelf-life in cold-storage of two to four weeks. Either form is best produced from fairly perfect, undamaged fruit to avoid extensive hand-trimming. An easily peelable shape and small seed pocket will help to minimize residual peel and carpel that are also undesirable and cause for down-grading under the U.S. Standards. Medium-size wedge (radial-cut) slices are preferred because large slices tend to be underblanched (with brown centers) and also to have excessive residual carpel attached.

When a sulfite antioxidant is used, there is likely to be faster, more complete penetration to the slice centers if the extent of voids in the tissue is great and the specific gravity therefore low. The tendency to browning discoloration should be slight in order to insure most acceptable color, especially in the lightly-sulfited, fresh slices. There is a premium on firm, cohesive texture that will encourage slice integrity throughout processing and later utilization in the bakery. Juiciness should be minimal to yield a high drained weight of frozen-thawed product or of the fresh product after refrigerated storage.

Canned slices. The requirements for canning apples are only slightly different from those described for fresh and frozen. Because of the severity of thermal processing in the usual solidly-packed #10 can, the apples have a greater need to be extra firm. This is often obtained by deliberate selection of hard varieties, harvesting at early maturity, or processing soon after harvest. Apples of high specific gravity would probably help in obtaining the required 7-lb fill, although this is partly a matter of the thermal or vacuum processing that replaces tissue gasses with liquid as completely as possible. Flooding of the tissue voids with processing medium may offset both low specific gravity and excessive loss of juice. A further reason for the replacement of gasses is to prevent early corrosion or failure by pin-holing of the can lining due to oxygen remaining in the product or head-space. Slices thoroughly cooked in a water medium will be especially bland in flavor, suggesting that an appreciably tart apple be used.

Dried or dehydrofrozen pieces. Dried apple is produced at two principal moisture levels: 18 - 24% for the traditional "evaporated apples" and 2 - 5% for the more fully dried product. Dehydrofrozen pieces are commonly dried to half their original weight and then preserved from microbial spoilage by freezing. These products include the increasingly popular diced form used in fried pies, turnovers, and other small bakery items. A high cut-yield of more uniform dices can be obtained with very large apples as long as they can be adequately machine-peeled. The smaller piece size common to these dried products is an advantage in that less thorough trimming of the peeled apple is necessary. Bruised tissue is more unsightly than unpalatable, so that the reduction of damaged areas to more innocuous bits by the smaller cut partly alleviates the need for trimming. This is obviously less true of decay than of simple bruises.

The use of a sulfite antioxidant is nearly universal, but because of the small piece size, sulfite penetration is not so critical as in slices. Hence, the greater tissue voids associated with low specific gravity may be less advantageous and perhaps counter-productive in permitting the SO_2 to be swept away during drying, leaving insufficient residual to insure effective antioxidant protection against browning, oxidative or otherwise, in the dried product during storage. A secondary sulfite treatment after drying and equilibration and before or during packaging is often practiced for that reason.

The greater the initial solids content in the fresh apple, the less drying needs to be done to reach the desired low moisture level. In a time of steadily increasing fuel cost, this can be a real advantage. On the other hand, a low to moderate acidity will also be concentrated by the drying process to a relatively intense level that may in part persist through the ordinarily incomplete (70%) reconstitution in the bakery product. This effect is probably the main reason why dried foods have an ill-deserved reputation for being especially tasty, whereas in truth much of the volatile flavor will have been irretrievably lost in drying.

A low rate of browning is helpful since heating the pieces to effect drying may follow close on the heels of antioxidant treatment - before it has had time to become fully effective. An older procedure used with the larger wedge-slices allowed some time for sulfite penetration to be completed before drying was started, but this led to excessive separation of solids-bearing juice from the tissue, the integrity of which is attacked by the sulfite. This 'weeping' from the cut surfaces, arising also from the natural juiciness of the apple, is anethema in the dryer, where it may cause excessive sticking of pieces to the conveying surface, besides reducing yield of dried product.

Applesauce. This distinctly North American product has only recently yielded first place in apple utilization to juice products. Various in its color, flavor, and texture, it enjoys favor among a wide public. The fact that not everyone likes the same style of sauce somewhat complicates the correlation of raw and processed product attempted in this paper. Still, certain elements are held in common.

Blending is usually, though not always, practiced both to improve sauce quality and to make it more uniform over the season of pack. Golden Delicious and McIntosh are made into single-variety sauces, perhaps not so much because they are especially good for the purpose as that they are well-known to the consuming public and available in large quantity, often as packing-house culls from the more well-deserved marketing of these varieties fresh. Gravenstein is also used in a single-variety sauce to somewhat better effect.

Unless an especially coarse or hard graininess of texture is desired, flavor is likely to be best in fully matured and well-ripened fruit. Quite a bit of the volatile flavor that maturation and ripening develop is lost in cooking by a sort of steam-distillation of the pulp. This involves a heat transfer-efficient direct contact of steam with finely-cut apple, followed by centrifugal pulping through a screen while still very hot. However, a well-made sauce should retain some aroma if these operations are effectively enclosed.

Decay must be removed by hand-trimming, but much of the bruised tissue and carpel will be eliminated as waste during pulping and finishing through successive fine screens (1/8-inch, or less). Red skin would impart its unstable color to sauce during cooking, so that commercial sauce is rarely made from unpeeled fruit. Since much of the superficial bruised tissue arising from normal harvest and handling is removed by peeling, even yellow- or green-skinned fruit is customarily peeled. Cores are also removed, not because of concern with carpel so much as to eliminate troublesome broken seeds and dark fragments from the blossom end. Hence, yield in this instance is sacrificed in the interests of quality.

Sauce color is usually subject to strong preference on the part of the buyer or in a particular market - whether yellow, white, or green-tinged. The blend of apple varieties may be chosen accordingly. There is considerable latitude in Brix-acid ratio ... anywhere from 25 to 60, and it is adjustable within limits by the addition of a variable level of sugar, or even of food acid. A recent move to marketing applesauce at lower Brix (15°) than the usual 18 - 20° range is partly in response to diet-conscious consumers and partly, as at present, because of the price of sugar rising above that of apple. To preserve the usual Brix-acid ratio in that case, apples of lower natural acidity should be used, which is easier to do rather than more difficult in some regions. The rate of browning is important in sauce processing, as a light, bright color is desired. Heating may be slow enough in some cookers that the enhanced enzyme activity experienced in the range 125 to 160°F, part way through the come-up to a cook temperature of 215°F, may permit rapid browning before enzyme inactivation is accomplished.

The amount of free liquid in applesauce is adjusted to provide the proper flow characteristics and mouthfeel. In some varieties, particularly when quite ripe, the natural separation of juice from the broken cell structure together with the steam incorporated as condensate (about 15% w/w) is more than enough, while to others water must be added with the sugar, increasing case yield in proportion. Apple varieties for sauce are sometimes priced accordingly.

Cider and juice. These products are frequently not distinguished in the trade, though some do insist on a difference. The freshly-pressed juice of the apple is naturally more or less cloudy as well as somewhat thick or syrupy in body, and it quickly oxidizes both in color and flavor. These attributes characterize cider in the older tradition. Enzyme-treatment and filtration that clarify the cider, and incidentally degrade the pectin and thin its body, render it apple juice, in my estimation. The same may be said of the use of antioxidant such as ascorbic acid to maintain the original light color, or pasteurization that adds a cooked note to the flavor. But, depending on how seriously you view the intrusion of chemical preservatives such as potassium sorbate or sodium benzoate into the flavor, the definition of cider might be bent a little to include their use.

Decay must be thoroughly removed by sorting and vigorous washing. Apart from aesthetics and the probable introduction of off-flavor, microbial spoilage of the cider or juice may be hastened by the failure to clean up the raw product. There is also a toxicological consideration in the possible occurrence of mold-induced patulin (13).

One would think there might be some correlation between the sensation of juiciness on biting into an apple and the ease with which a normally high press-yield of 80 per cent or more could be obtained. That may be so, but the data to prove it seems scarce. This may be a case of "back to the laboratory"...

The rate and ultimate degree of browning have least effect on the preparation of a "natural" light-colored juice. Since the ascorbic acid can be added (typically at .05%) directly into the mill as the apple is finely chopped before pressing, there is little difficulty in controlling enzymatic browning quite completely. There is even reason to think that the trade will more readily accept such a juice if a delay in adding the ascorbic acid is practiced to permit a little irreversible browning to develop a more recognizable light amber or golden color. The desired delay would depend on juice temperature, oxygen contact, and the intrinsic potential of the apple variety for browning.

A high level of suspended solids in cider reduces its translucency, possibly making it appear unattractively muddy and contributing to early and excessive formation of sediment in the consumer package. This might be considered an artifact as much as a factor of raw product quality, since milling and pressing methods affect it so directly. Overripe, out-of-condition fruit - such as is very often used - is a principal contributing factor. While the suspended solids may be completely removed from clarified juice, filtration efficiency and final juice yield would be adversely affected by their presence in excess of one or two per cent. Hazes appearing in clarified juice are sometimes a problem; their chemistry has been recently discussed (14, 15).

Although the flavor of the apple is carried over into cider and juice, at least in terms of sugar/acid balance, little deliberate selection of raw product is practiced on that account. What is available economically is what is used for these products. But since blending of different varieties and lots of apples is usual, product quality remains near some acceptable if undistinguished mean. In general, an apple right for eating out of hand is also right for cider or juice, remaining a somewhat subjective judgment in that sense.

Juice concentrate. This product has become an important commodity in world trade, its many uses not limited just to reconstitution into single-strength juice. The normal 7:1 (72° Brix) concentrate is convenient and economical to store and ship, needing refrigeration only to reduce usual browning. Its requirements for raw product quality are substantially the same as for single-strength juice. Since brown discoloration is itself concentrated in the product and added to by possible caramelization in the evaporator and non-enzymatic browning during long storage at ambient, it is even more important to minimize enzymatic browning during the initial processing. As in dried apple, concentration places a premium on high starting solids, though in this case only solubles are involved.

Hard cider and wine. While apple variety is considered important in these apple products, it may be mainly a matter of tradition. True, the British and European cider apples are (or more accurately, were) special sorts, remarkable as well for their mean size and unpalatability fresh as for the pronounced astringency that provides a distinctive cider flavor. However, the North American market for a low-alcohol (6%) hard cider or apple wine is so relatively undeveloped that product quality is not yet well-defined. Hence, some popular dessert apples - even, or especially, the slightly astringent and low-acid Red Delicious - may be quite adaptable. Among the few requirements that can be clearly seen at present are fully mature and well-ripened fruit (perhaps to the point of declining pressability), the lower acidity associated with it, and at least some pretense of astringency to add a desirable complexity to the flavor sensation. The choice of carbonated or sweetened versions of the product, as opposed to still or dry, would probably reflect differently on raw product specifications.

Apple wine, legally defined as having greater than 7 per cent alcohol by volume, would benefit from high soluble solids to reduce the need for added sugar. A pronounced fruitiness might carry over into the finished wine to better advantage than in hard cider, lest the greater alcohol content predominate in flavor. As in sweet cider and juice, the raw product requirements of these fermented beverages may not be difficult to meet, product quality being more certainly dependent on good manufacturing practices.

Other products. Of the remaining products listed in Table 1, less needs to be said. Except for vinegar, which is largely made from peel-and-core waste anyway, they are all minor uses of apple. Not much apple character is retained in such thoroughly transformed products as vinegar and apple butter. A reasonable substitute for the former is even made from synthetic ethanol, as well as from other fruits, while the latter is not only given a lengthy caramelizing cook but is heavily spiced as well. In each, however, a high level of fruit sugar would be advantageous - as the fermentable substrate for acetic acid in the first and as a headstart on the desired concentration and thickening of consistency in the other.

The principal requirement for spiced apple rings is a firm apple texture that will perform well as a carrier for the added color and flavor, much the same as in maraschino cherries, though uniform fruit size may also be a factor. Apple jelly could presumably be made without additives other than sugar if pectin, acid, and pigmentation were at the required levels, as in crabapples. Apple essence, derived from stripping volatiles from depectinized, single-strength juice prior to its concentration, is an undefined pot pouri of flavor elements chiefly specified in terms of its concentration or "-fold". Nectar or pulp, prepared as a single-strength or somewhat concentrated homogenates of whole apple, are not presently articles of commerce in the United States. However, some processes for and grades of applesauce have approximated this simplified approach, gaining in yield at the expense of quality.

Other considerations. Little has been said here of the role of the pectic substances, and the omission may be a glaring one in view of their certain involvement in the quality of many of these products. As in the polyphenols, the chemistry is complex and beyond the scope of this review. Pectin is intentionally removed from juice to be clarified by employing enzymatic degradation; this effects the desired colloidal change leading to agglomeration of particles and also prevents early clogging of filter surface by gelatinous deposits. Likewise in juice to be concentrated, pectin must be removed to avoid fouling of evaporator surface. In these instances its absence may be qualitatively tested for with alcohol, in which it is insoluble. Pectin is naturally degraded during the fermentation of hard cider and wine, resulting in the desired autoclarification.

However, pectin in the appropriate chemical form is necessary for good texture of apple pieces and of sauce, providing the cement that insures cell cohesion. Its presence may also reduce the separation of liquid from applesauce. The firming of piece texture with added calcium involves a beneficial pectic transformation (16). And, in the manufacture of jelly, pectin may have to be added to obtain the desired gel strength.

What is certain from this analysis is that raw product should be specified and purchased - perhaps even specially grown - to suit the particular needs of processing the intended product. That would be the shortest path to insure a quality product at least processing cost. To a large extent, and necessarily, this selection is practiced only with the crop or supply available at the time of need. The more critical is raw product quality to successful processing, e.g., for frozen slices, the greater has been the effort to meet the quality requirements by preliminary contract or arrangement. Some processing plants are large enough to process several products having different enough requirements to permit best use of available lots of apples. A typical example is the frequent utilization of small apples, that can't economically be peeled, in a juice line along with peel-and-core waste from sauce or slice processing. Although practiced to some extent by the industry as a whole, the selection process becomes inefficient when too many conflicting and competing decision-makers act independently, even though guided by pricing incentives. There is room for improvement in overall management of the crop, and in tailoring future supplies to changing needs. This remains an important concern of the industry.

Literature Cited

1. Johnson, J. M.,; Wood, C. B.; Mattus, G. E. V.P.I. & S.U. Ext. Div. Publ. 759. 1977.
2. Anon. IAI Special Letter 1980, 11(9).
3. La Belle, R. L.; Lind, L. R.; Kime, R. W. Proc. N.Y.S. Hort. Soc. 1978, 123, 180-183.
4. La Belle, R. L. Unpublished data. 1978.
5. Reeve, R. M. Food Res. 1953, 18, 604.
6. Magness, J. R.; Taylor, G. F. U.S.D.A. Circ. 350. 1925.
7. Blanpied, G. D.; Bramlage, W. J.; Dewey, D. H.; La Belle, R. L.; Massey, L. M., Jr.; Mattus, G. E.; Stiles, W. C.; Watada, A. E. New York's Food & Life Sci. Bul. No. 74. 1978.
8. La Belle, R. L.; Shallenberger, R. S.; Way, R. D.; Mattick, L. R.; Moyer, J. C. Food Technol. 1960, 15(9), 463-468.
9. Plane, R. A.; Mattick, L. R.; Weirs, L. D. Amer. J. Enol. 1980, 31(3), 265-268.
10. Williams, A. A. J. Inst. Brew. (London). 1974, 80, 455.
11. Lea, A. G. H. J. Sci. Food & Agric. 1978, 29, 471-477.
12. Williams, A. A.; Lea, A. G. H.; Timberlake, C. F. In: "Flavor Quality: Objective Measurement". ACS Symposium Series 51. Amer. Chem. Soc., Washington, DC 1977, 71-88.
13. Wilson, D. M.; Nuovo, G. J. J. Appl. Microbiol. 1973, 26, 124-125.
14. Heatherbell, D. A. Alimenta. 1976, 15, 151-154.
15. Van Buren, J. P. Voedingsmiddelentechnol. 1972, 3(13), 67.
16. Collins, J. Ph.D. Thesis. Univ. of Md., College Park. 1966.

RECEIVED April 3, 1981

7

Establishing Criteria for Determining the Authenticity of Fruit Juice Concentrates

RONALD E. WROLSTAD[1], CHRISTOPHER J. CORNWELL[2], JEFFREY D. CULBERTSON[3], and FELIX G. R. REYES[4]

Department of Food Science and Technology, Oregon State University, Corvallis, OR 97331

Use of fruit juice concentrates in the manufacture of wines, jellies, and other foods and beverages is increasing and fruit juice concentrate production has become a highly competitive growth industry in both domestic and foreign markets. Analytical methods for determining the authenticity of fruit juice concentrates are needed for quality assurance and ingredient specification as well as for regulatory activities. Consumers have become aware of possible adulteration problems through investigative journalism (1, 2). By C^{12}/C^{13} analyses many apple products purchased from Boston and New York supermarkets were shown to be adulterated with corn syrup or cane sugar. This sophisticated analytical technique has proven very useful for detecting adulteration of honey with corn syrup and cane sugar. It cannot be used, however, for detecting addition of beet sugar or sugars derived from other plants that utilize the Calvin pathway for CO_2 fixation (3, 4).

Fruit juice concentrates vary widely in price because of differences in raw product cost and processing yield. For example, wholesale 1978 prices for Pacific Northwest produced concentrates varied from $6.25 per gallon for apple and pear to $35, $75, and $110 per gallon for strawberry, blackberry, and black raspberry concentrates, respectively (5). The readily available, low priced concentrates are obvious potential adulterants for more expensive concentrates and methods are needed for their detection.

Factors such as variety, maturity, climate, geographic origin, cultural practices, and the influences of stress and disease, can affect a plant's chemical composition. Changes in composition can occur during processing and storage and certain additives are

[1] To whom correspondence shall be addressed

[2] Current address: Department of Food Science, Cornell University, Ithaca, NY 14853

[3] Current address: Department of Food Science and Technology, Washington State University, Pullman, WA 99164

[4] Current address: Faculdade de Engenharia de Alimentos e Agricola, UNICAMP, Campinas, S. P. Brazil 13100

0097-6156/81/0170-0077$05.00/0

sometimes permitted as processing aids which confound the proof of authenticity by compositional analyses. We attempted to consider many of these factors in a previous study (6) of sugar and non-volatile acid composition of 14 blackberry samples and while this study was relatively small in scope it still required considerable effort. We wanted to obtain reference compositional data for fruits of major economic importance to the Pacific Northwest - grape, apple, pear, strawberry, plum, blackberry, cherry, and raspberry. Analysis of an adequate number of samples to include varietal, geographical, seasonal, and maturity differences for these fruits would be a formidable task. Tabulation of authentic data from the literature is necessary to indicate priorities for further compositional investigations. Such a compilation should begin to indicate variability due to variety and geographic origin and would also include variation caused by different methods of analysis.

Free Sugars and Sorbitol

A computer search for the glucose, fructose, sucrose, and sorbitol contents of apple, pear, cherry, plum, grape, blackberry, raspberry, strawberry, and peach fruit was conducted and the data were compiled both as g/100 g fresh weight and as percent of total sugars. Total sugars and glucose:fructose ratios were calculated and the range, mean, standard deviation, and percent coefficient of variance determined for all entries. Thirteen tables resulting from this compilation of data were published as a separate paper (7) which can be referred to for specific information; Tables I and II are derived from those 13 tables and reveal the comparative sugar and sorbitol patterns for the different fruits and their variability. Considering the sample diversity and the different analytical methods used, one might expect to find even greater variation in the data than was obtained. The data sources for apple fruit, for example, represent several different varieties from Western and Eastern USA, Canada, Germany, Switzerland, England, Italy, and New Zealand and a wide variety of analytical methods - gas liquid chromatography (GLC), paper chromatography, ion exchange chromatography, enzymic and wet chemical analyses. The percent coefficient of variance (% C.V.) for fructose expressed as percent of total sugars is only 8.8. The % C.V. for fructose content expressed on a g/100 g basis is 9.8. The % C.V. for the individual sugars calculated as percent of total sugars is lower than the % C.V. for the same sugars reported on a g/100 g fresh fruit basis. This holds true for all fruits and is not surprising since sugar determination on a g/100 g fresh weight basis will reflect both variation in total solids and analytical errors in quantitative recovery of total sugars. Dako et al. (8) analyzed 10 to 12 samples for each of the following fruits over a period of two years and reported the following ranges for total sugar content (g/100 g): apple, 6.03-15.25; pear, 5.77-9.74; plum, 6.33-10.83; cherry, 9.51-

Table I. Average Fructose and Glucose Contents of Fruits[a]

Sample	n	Fructose		Glucose		Glucose:Fructose	
		% T.S.	% C.V.	% T.S.	% C.V.	glu:fru	% C.V.
Apple	13	55.0	8.8	21.0	32.2	0.37	38.2
Processed apple	13	66.3	19.4	23.0	20.0	0.37	34.3
Pear	16	65	13.0	16.8	3.8	0.26	37.8
Processed pear	5	57.9	29.0	23.9	51.9	0.48	72.7
Plum	13	23.0	49.1	38.3	28.7	1.94	42.4
Processed plums	8	38.1	21.5	60.8	13.0	1.69	33.2
Sweet cherry	11	46.4	16.0	52.3	15.4	1.17	30.5
Sour cherry	5	44.0	15.8	52.5	5.6	1.22	20.6
Grapes	8	45.0	12.4	46.5	16.9	1.03	9.3
Strawberry	9	40.4	19.6	42.5	22.8	1.16	61.2
Red raspberry	6	34.6	29.8	31.3	31.7	0.94	33.0
Blackberry	6	48.0	15.2	46.8	10.4	1.00	17.4
Peach	8	12.4	35.8	11.9	17.9	1.30	91.5

[a] Derived from compilations from the literature (7); sugar content listed as percent total sugars (T.S.); % C.V. = percent coefficient of variance; n = number of sources; glu:fru = glucose:fructose ratio.

Table II. Average Sucrose, Total Sugar, and Sorbitol Contents of Fruits[a]

Sample	Sucrose			Total Sugars			Sorbitol		
	% T.S.[b]	n	% C.V.	g/100g	n	% C.V.	% T.S.+S.[c]	n	% C.V.
Apple	24.1	14	31.9	11.0	12	14.8	4.74	8	57.2
Processed apple	10.5	13	92.2	33.8	13	59.8	4.16	8	66.1
Pear	16.8	19	50.0	10.2	18	11.8	17.6	13	24.8
Processed pear	16.7	5	55.3	30.9	5	81.2	13.2	4	51.4
Plum	38.4	13	49.5	8.79	12	26.3	16.0	5	39.0
Processed	1.22	8	184	46.3	8	22.5	24.3	6	17.4
Sweet cherry	1.04	11	160	15.1	8	29.8	16.9	4	41.7
Sour cherry	3.46	5	143	8.71	5	25.4		0	
Grapes	3.58	8	112	14.7	8	19.6	0.6	3	
Strawberry	16.6	9	39.8	5.5	9	19.6	0	1	
Red raspberry	33.6	6	16.2	5.82	6	11.7		0	
Blackberry	3.7	6	123	6.96	6	17.4	0	3	
Peach	75.7	8	7.3	8.12	7	14.0	7.17	3	82.0

[a] Derived from compilations from the literature (7); n = number of sources; % C.V. = percent of coeffecient of variance

[b] % total sugars

[c] % total sugars + sorbitol

16.06; peach, 4.65-8.58; grapes, 8.75-17.30; strawberry, 3.96-5.96; raspberry, 3.66-9.30; blackberry, 4.78-7.38.

Patterns characteristic of the individual fruits are evident from Tables I and II. Pear and apple contain much more fructose than glucose while raspberry, blackberry, grape, strawberry, and cherry have essentially invert sugar (glucose:fructose = 1:1) patterns. Conversely, peach and plum contain more glucose than fructose. The glucose:fructose ratio is useful, therefore, in categorizing fruits. It will be altered, however, by sucrose inversion in fruits not having invert patterns. Kliewer (9) measured the glucose:fructose ratio in eight varieties of grapes during ripening and found that the 1:1 ratio in ripe fruit changes to 0.63 in overripe fruit.

Table II shows that sucrose levels are highest in peach fruit and very low in blackberry, cherry, and grape. The % C.V. for sucrose content for many of the fruit samples is extremely high, most likely because of enzymic or chemical hydrolysis. The average sucrose content of plum fruit is 38% of total sugars, while processed plums contain only 1%. Presence of large quantities of sucrose in prune juice has, in fact, been used as an indication of adulteration (10). Cherries and grapes contain invertase (11, 12) which may account for sucrose not being detected in some samples and may also explain the high % C.V. for those fruits. In examining apple juice concentrates, we did not detect sucrose in seven-fold concentrates whereas we did in four-fold apple concentrates (7), the time-temperature conditions allowing for complete hydrolysis in the more concentrated product.

Sorbitol content can be a useful index in classifying fruits. Pear, cherry, and plum fruit are highest in sorbitol and both peach and apple contain substantial quantities (Table II). Trace amounts were found in grapes (13); no sorbitol was reported for blackberry, raspberry, and strawberry (Table III). The latter three fruits are of high economic value and the presence of sorbitol in their concentrates can be an indication of adulteration with less expensive sorbitol-containing fruits. The % C.V. for sorbitol content is fairly high for some fruits. In some instances, e.g. peach, the relatively small sample size may be a factor. Neubeller and Stosser (14) studied the sugar and sorbitol variation of 20 varieties of sweet cherries over three years and reported the sorbitol content to vary from 15.5 to 24.0 percent of total sugars plus sorbitol. Chan *et al* (15) showed that sorbitol content of apples can show considerable variation with variety, storage conditions, and season.

We examined a number of commercial blackbery juice concentrates by GLC of their trimethylsilyl (TMS) derivatives. Three samples were found to contain from 2.8-21.5% of total sugars as sorbitol (16) leading us to believe that those samples were adulterated. Previously, we had examined 14 authentic samples of blackberries by similar analytical procedures and no sorbitol was detected in any of those samples (6). Sorbitol is non-fermentative and when we made wine from sorbitol-containing blackberry juice concentrate we readily detected it by high performance liquid chromatography (HPLC). We looked at three commercial wines, and one which was labeled as being produced from imported European blackberries contained sorbitol. Two other domestic wines may have contained trace amounts of sorbitol because a small peak with the same retention time as sorbitol was detected. Makinen and Soderling (17) recently reported that ripe strawberry and ripe red raspberry contain 320 and 85 μg sorbitol/g fresh fruit, respectively. Previous workers have not found sorbitol in strawberries and raspberries; future work using more sensitive analytical methods may show that blackberries and other berry fruits also contain sorbitol, if only in very small amounts. If the raw product from which a juice concentrate was made had been partially fermented, this could result in increased concentration of any sorbitol which might allow its detection. Analysis of commercial apple juice concentrates in our laboratory revealed that the sorbitol content of one sample was far below the range for sorbitol content reported in the literature. That sample had a high glucose:fructose ratio leading to the suspicion that it may have been adulterated with corn syrup. This illustrates how sugar analysis can be a useful screening mechanism but the suspect sample should be subjected to further analyses for more definitive proof.

Nonvolatile Acids

Table III lists the malic, citric, isocitric, lactoisocitric, tartaric, quinic and total acid contents for apple, pear, plum, cherry, grape, strawberry, raspberry, blackberry, peach, and cranberry fruit as reported in selected literature references. A wide range of values for total acids occurs both within and between the different commodities, limiting its utility in detecting adulteration. This is not surprising because of the abrupt drop in total acidity which will occur with many fruits when reaching the fully ripe stage, and the known influences of variety and climatic conditions on acidity (18). There are several contradictions in Table III as to the qualitative composition of a given fruit. For example, quinic acid is reported for some but not all samples of plum, apple, grape, strawberry, and peach fruit. Benk (19) reported that strawberry juice contains isocitric as well as citric acid and that raspberry juice contains more isocitric than citric acid. Whiting (20) reported that blackberries contain isocitric and not citric acid while Benk (19) reported that blackberries contain both with isocitric being present in larger quantities than citric acid. We were unable to resolve citric and isocitric acid in our analyses of blackberry fruit by GLC (6). These two acids are difficult to resolve by GLC and citric acid content may be higher than the correct value in reports for other fruits because of presence of isocitric acid. Fernandez-Flores *et al.* (21) reported syringic acid to be a major acid in cherrries and also reported its presence in plums, apples, and pears but this has not been confirmed by other workers. Phosphoric acid apparently can be found in small quantities in most fruits when the analytical methods permit its detection. Ryan and Dupont (22) pointed out that the GLC retention time for the trimethylsilyl (TMS) derivative of phosphoric acid is very similar to those for succinic, fumaric, malonic, and maleic acids on both SE-30 and OV-17 liquid phases; these acids as well as glycolic, shikimic, and chlorogenic have previously been reported in small quantities in many fruits (18). They are not listed in Table III because their detection in a fruit concentrate is of limited value for adulteration analyses. With increased resolution and sensitivity in methods for nonvolatile acid analyses, it is likely that they will be detected in authentic samples of many fruits in small quantities.

There is considerable variation in the relative amounts of the major acids in Table III for most of the fruits. For example, one reference shows malic acid to be the major acid in peaches while another shows citric to be the principal acid. In analyzing the changes of nonvolatile acids during the ripening of Bartlett pears, we found malic to be the major acid in underripe fruit and citric to be present in largest quantities in ripe fruit (23). Cash *et al* (24) found malic to be present in larger quantities than tartaric in less mature Concord grapes, but tartaric was present in relatively larger amounts that malic in the fully ripe fruit. Johnson and Nagel (25) similarly found tartaric in greater quantity than malic acid in five varieties of grapes when the fruit was fully mature. We found a great deal of quantitative variation in blackberry nonvolatile acids, isocitric being the principle acid in some samples, malic being the predominant acid in others, and lactoisocitric being the major acid in a very underripe sample of Evergreen blackberries (6).

Some acids are characteristic of individual fruits and quite useful in adulteration investigations. Tartaric acid is a major acid of grapes and is not found in large quantities in the other fruits. However, removal of tartaric acid through crystallization and ion exchange procedures can result in its being present in only minor quantities in grape juice and concentrates. Hence, it may be present in only minor amounts in a fruit concentrate adulterated with grape juice concentrate. Lactoisocitric acid has a limited distribution in fruits, but is found in blackberries. Our results showed that it is not present in all blackberry samples, however, and its absence in blackberry juice will not deny its authenticity (6). Quinic acid is quite widespread among fruits and is present in quite large concentrations in plum and cranberry. We found high levels of quinic acid in an imported blackberry juice concentrate which also contained sorbitol; this led us to suspect that the sorbitol-containing adulterant may have been plum (16).

Acids such as citric and malic are permitted for use in pH adjustment in fruit juice processing which may confound the inter-

Table III. Major Nonvolatile Acids in Fruits

Sample	n[a]	Malic % T.A.	Citric % T.A.	Isocitric % T.A.	Lactoisocitric % T.A.	Tartaric % T.A.	Quinic % T.A.	T.A.[b] mg/100g	Ref.
Apple									
apple juice	6	91.3	2.0				4.0	701	(39)
Granny Smith		69.3	22.3				5.6	548	(40)
Delicious		35.7	17.2					568	(21)
Winesap		60.1	10.3					1085	(21)
Pear									
Bartlett		54.4	18.7				7.8	535	(21)
juice	1	79.6	3.6				6.8	210	(23)
Plum									
prune juice	1	21.6					78.4	791	(39)
Italian		88.3					10.0	2203	(21)
plum		85.5	12.4					1814	(21)
concentrate (10° Brix)	2	55.7	1.6				30.7	515	(41)
Cherry									
fruit		25.1	18.9					243	(21)
Grape									
juice	4	45.7				51.5		1029	(39)
red		37.6	13.6			34.5	9.8	1135	(21)
concord		51.6	5.7			37.9	4.8	1759	(21)

Table III (continued)

Sample	n[a]	Malic % T.A.	Citric % T.A.	Isocitric % T.A.	Lactoisocitric % T.A.	Tartaric % T.A.	Quinic % T.A.	T.A.[b] mg/100g	Ref.
Strawberry									
juice	1	35.4	59.5					975	(39)
fruit		7.9	85.6				2.8	1098	(21)
fruit (full ripe)	4	20.6	73.6				1.3	1220	(42)
Raspberry									
juice	1		97.4					2547	(39)
Malling Promise		4.1	95.9					1968	(20)
Blackberry									
fruit	14	15.0		59.9[d]	13.7		0.7	1049	(6)
fruit		47.6		17.5	34.9		tr[c]	1718	(20)
Peach									
fruit		48.9	32.2				17.9	994	(21)
fruit	3	35.7	63.5					928	(43)
Cranberry									
juice	1	28.5	32.0				39.5	632	(39)
juice	27	27.7	32.5				39.8	3320	(26)

[a] n = indicated sample size

[b] T.A. = total acids

[c] Tr = trace

pretation of nonvolatile acid profiles of fruit juices and concentrates.

Benzoic acid is found naturally in cranberries (0.6% of total acids; (26)). In other fruits its presence would indicate addition of sodium benzoate.

Free Amino Acids

Table IV lists the proline, arginine, aspartic acid, glutamic acid, aminobutyric acid, alanine and total amino acid contents of several fruits as reported in selected literature references. A particularly salient reference is that of Fernandez-Flores *et al* (27) who reported quantities of 15 free amino acids for 22 different fruits. They suggested that the qualitative and quantitative distribution of free amino acids could be characteristic of the individual fruits. There were only small variations in the patterns for six different orange samples and seven strawberry samples. They concluded that the qualitative and quantitative distribution of amino acids could be a useful index for determining the authenticity of fruit products. Also, total amino acid content itself can be effectively used since apple and cranberry are very low in total amino acids while grape, blackberry, plum, and peach are relatively high.

Colorimetric determination of proline content can be a useful quality control measurement. Grapes are fairly high in proline, and apples are quite low. The U.S.A. import duty for grape juice concentrate is $0.25 per gallon on a single strength basis whereas apple and pear juice concentrate are duty-free (28). Measurement of proline content is helpful in determining entry of grape juice concentrate under an illegal classification.

Swiss law permits amelioration of apple juice with pear juice to a maximum of 10 percent. Blumenthal and Hebling (29) used a combination of proline, citric acid, and sorbitol analyses to establish base-line reference data for determining whether the permitted 10 percent amelioration had been exceeded. They found that one third of the commercial apple juices investigated did not comply with the legal requirements. Their approach to solving their difficult analytical problem illustrates the necessity of using a combination of parameters rather than relying on measurement of one particular constituent. Other amino acid determinations which may have useful application relate to the high alanine content of blackberries, the high aminobutyric acid content of grapes and blackberries, and the high aspartic acid content of pears and plums.

Attention needs to be given to changes in free amino acid content which occurs during processing and storage. Wucherpfennig (30) showed that a marked reduction in proline and arginine occurred in processing grape juice concentrate. Lowering amino acid content with cation exchange resin treatment is permitted by US law in non-standardized fruit juices and is used to reduce

Table IV. Principle Amino Acids Useful in Characterizing Fruits[a]

Sample	n[b]	Proline mg/100 g	Arginine mg/100 g	Aspartic mg/100 g	Glutamic mg/100 g	Aminobutyric mg/100 g	Alanine mg/100g	Total mg/100g	Ref.
Apple									
fruit		tr	n.a.	0.6	0.9	0.3	1.0	3.3	(27)
juice		0.6	tr	n.a.	n.a.	n.a.	n.a.	n.a.	(30)
juice	7	0.37	n.a.	n.a.	n.a.	n.a.	n.a.	n.a.	(29)
juice	9	n.a.	n.a.	n.a.	n.a.	n.a.	n.a.	0.308[c]	(44)
Pear									
fruit		1.8	n.a.	23.3	1.3	tr	0.5	31.8	(27)
juice (Bartlett)	3	6.0	0.6	4.5	3.3	n.a.	1.4	75.5	(45)
juice		23.0	n.a.	n.a.	n.a.	n.a.	n.a.	n.a.	(30)
juice		22.8	n.a.	n.a.	n.a.	n.a.	n.a.	n.a.	(29)
Plum									
fruit (purple)	2	2.6	n.a.	96.1	3.7	4.6	1.7	110	(27)
Fruit (yellow)		49.7	n.a.	184.9	7.9	33.5	24.8	327	(27)
Cherry									
Sour juice		13.6	n.a.	n.a.	n.a.	n.a.	n.a.	n.a.	(19)

Table IV. (continued)

Sample	n[b]	Proline mg/100 g	Arginine mg/100 g	Aspartic mg/100 g	Glutamic mg/100 g	Aminobutyric mg/100 g	Alanine mg/100g	Total mg/100g	Ref.
Grape									
black		4.4	n.a.	6.5	43.6	19.1	13.4	135.4	(27)
red		39.7	n.a.	8.9	36.6	14.7	5.9	123.6	(27)
white		39.1	n.a.	2.1	35.0	22.4	32.2	172.8	(27)
Fruit		30	400-1800	n.a.	n.a.	n.a.	n.a.	n.a.	(30)
Ripe fruit	18	125	108	13	51	22	35	398	(46)
Strawberry									
fruit	7	0.8	n.a.	30.4	12.4	5.6	5.8	62.5	(27)
Hood	1	tr	1.1	4.5	9.7	n.a.	5.4	62.5	(47)
Tioga	1	tr	0.6	4.3	7.3	n.a.	5.7	55.1	(47)
Juice		1.2	n.a.	n.a.	n.a.	n.a.	n.a.	n.a.	(19)
Fruit	17	n.a.	n.a.	n.a.	n.a.	n.a.	n.a.	0.82[c]	(48)
Fruit	77	n.a.	n.a.	n.a.	n.a.	n.a.	n.a.	1.05[c]	(39)
Raspberry									
Juice (red)		tr	n.a.	n.a.	n.a.	n.a.	n.a.	n.a.	(19)
Juice (red)	34	5.95	n.a.	n.a.	n.a.	n.a.	n.a.	n.a.	(49)
Fruit (red)	16	n.a.	n.a.	n.a.	n.a.	n.a.	n.a.	2.14[c]	(48)
Fruit (black)	3	n.a.	n.a.	n.a.	n.a.	n.a.	n.a.	1.42[c]	(48)
Fruit	34	n.a.	n.a.	n.a.	n.a.	n.a.	n.a.	2.19[c]	(39)

Table IV. (continued)

Sample	n[b]	Proline mg/100 g	Arginine mg/100 g	Aspartic mg/100 g	Glutamic mg/100 g	Aminobutyric mg/100 g	Alanine mg/100g	Total mg/100g	Ref.
Blackberry									
Fruit		5.2	n.a.	56.2	17.4	25.1	80.9	228.9	(27)
juice		3	n.a.	n.a.	n.a.	n.a.	n.a.	n.a.	(19)
Fruit		n.a.	n.a.	n.a.	n.a.	n.a.	n.a.	2.25[c]	(48)
Peach freestone		tr	n.a.	105.6	1.6	0.8	0.8	111.9	(27)
Cranberry fruit		tr	n.a.	6.2	2.1	3.9	tr	15.1	(27)

[a] Values are mg/100 g fruit or 100 ml juice; n.a. = not analyzed; tr = tr

[b] Indicated sample size

[c] meq/100 g

color degradation through Maillard browning reactions in manufacture of pineapple juice concentrate. This severely restricts the utility of amino acid analyses in certain adulteration investigations.

Anthocyanin Pigments and Other Secondary Plant Metabolites

The anthocyanin pigment composition for many fruits is quite distinctive and is useful in certain cases of adulteration investigation. There are a number of reviewers who have summarized the anthocyanin pigments which have been identified in various fruits (31, 32, 33). Paper chromatography of anthocyanin pigments has been used to detect adulteration of Concord grape juice with Vinifera or hybrid varieties (34, 35) and to detect adulteration of blackberry and cherry juice concentrates with elderberry or grape skin extract (36). As mentioned before, we suspected that an imported blackberry juice concentrate sample was adulterated with plum juice because of its sorbitol and quinic acid content (16). Plums contain peonidin 3-glucoside and rutinoside in addition to the cyanidin-3-glucoside and cyanidin-3-rutinoside, which have been identified to be the pigments of blackberries (37). We failed to find any peonidin pigments in that imported blackberry concentrate sample. This did not prove that plums were not the adulterant because plums have relatively low concentrations of anthocyanins which are also unstable and may have been degraded. Pigment analyses of some blackbery concentrate and wine samples which we believed to be authentic, did reveal the presence of some pigments in addition to cyanidin-3-glucoside and cyanidin-3-rutinoside. These pigments may be minor compounds which have not been previously identified. It has been our experience that when we have fractionated large amounts of pigments through preparative column chromatography we invariably have found new pigments which had not been detected before (38). These pigments may possibly be artifacts of analysis or, in the analysis of wines and juice concentrates, may have been degradative products formed during processing and storage. Therefore caution must be used in interpreting anthocyanin data in adulteration studies. Analysis of anthocyanins by HPLC should give more meaningful quantitative data which would be easier to interpret than the qualitative patterns obtained in paper and thin-layer chromatography. The shorter analysis time using HPLC would also be more amenable to screening a larger number of samples. Analysis of flavonoids other than anthocyanins may prove useful since many of these compounds are more stable than the labile anthocyanins. There is a need for more definite identification work on the flavonoid composition of many fruits of economic importance.

In our investigations to date we have not used analyses of fruit juice concentrate volatiles to detect adulteration. The majority of the volatiles are removed in the concentration process, and the recovered essence may be added back to the concentrate or sold separately. Many of the less expensive concentrates are replacements for sugar syrups and are very bland in flavor; they are the most likely candidates for use as adulterants and would probably be difficult to detect by volatile analysis. However, there may be cases where analysis of volatiles would be useful in identifying an adulterant.

Conclusions

Detection of adulteration in fruit juice concentrates is a difficult problem because of compositional variation induced by varietal, maturity, seasonal, and geographic factors as well as the changes which occur during processing and storage. Addition of additives during processing and regulatory differences between countries add to the complexity of the problem. Still, there are patterns in the sugar, nonvolatile acid, amino acid, and pigment composition which are characteristic of the individual fruits and can be used to detect adulteration. It is undoubtedly true that our knowledge of food composition lags behind our present analytical capabilities. There is a particular need to further extend the knowledge of changes in food composition which occur during processing and storage.

Acknowledgements

Technical Paper No. 5606 from the Oregon Agricultural Experiment Station. The work was supported in part by grant No. 10790054 from the Pacific Northwest Regional Commission.

Abstract

The glucose, fructose, sorbitol, and sucrose contents of apple, pear, grape, cherry, plum, peach, strawberry, blackberry, red raspberry, and black raspberry fruit were compiled from the literature and analyzed statistically. The different fruits show characteristic sugar patterns with only moderate variation when considering the differences in geographic origin, variety, maturity, and method of analysis. Investigations in our laboratory utilizing GLC and HPLC reaffirm that sorbitol content and the glucose:fructose ratio are useful screening indices. Sucrose content shows considerable variation due to invertase activity and/or chemical hydrolysis, particularly in processed products. There is considerable quantitative variation in nonvolatile acid composition due to variety and maturity effects. Certain qualitative differences such as tartaric acid in grapes and lactoisocitric in blackberries are of practical use in detecting adulteration in certain cases. The free amino acid and anthocyanin pigment patterns are characteristic of individual fruits but changes in their composition which occur during processing and storage confound their utility in adulteration investigations.

Literature Cited

1. Beaton, A., Gold, M., Boston Herald American, Feb. 11, 1979, p. A1.
2. Wells, P., The New York Times, October 31, 1979, p C1.
3. Doner, L. W., Kushnir, I., White, J. W., Anal. Chem. 1979, 224A.
4. Doner, L. W., Krueger, H. W., Reesman, R. H., J. Agric. Food Chem. 1980, 28, 362.
5. Wrolstad, R. E., in "Fruit and Berry Concentrate Processing in the Pacific Northwest", Ninth Quarterly Report to Pacific Northwest Regional Commission. 1979. Dept. of Food Science & Technology, Corvallis, OR., p. 9.
6. Wrolstad, R. E., Culbertson, J. D., Nagaki, D. A., Madero, C. F., J. Agric. Food Chem, 1980, 28, 553.
7. Wrolstad, R. E., Shallenberger, R. S., J. Assoc. Off. Anal. Chem. 1981, 64, 91.
8. Dako, D. Y., Trautner, K., Smogyi, J. C., Schweiz. med. Wschr. 1970, 100, 897.
9. Kliewer, W. M., Amer. J. Enol. Viticult. 1965, 16, 105.
10. Flynn, C., Wendt, A. S., J. Assoc. Off. Anal. Chem. 1970, 53, 1067.
11. Glinskaya, Z. I., Shkurina, N. I., Ferment. Spirt. Prom. 1969, 35, 29; Food Sci. & Techno. Abstr. 1970, 2, 3J255.
12. Selvaraj, Y., Suresh, E. R., Divakar, N. G., Randhawa, G., Negi, S. S., J. Food Sci. & Technol. India 1975, 12, 75.
13. Kline, D. A., Fernandez-Flores, E., and Johnson, A. R., J. Assoc. Off. Anal. Chem. 1970, 53, 1198.
14. Neubeller, J., Stosser, R., Gartenbauwissenschaft 1977, 42, 222.
15. Chan, W. W., Chong, C., Taper, C. D., Can. J. Plant Sci. 1972, 52, 743.
16. Wrolstad, R. E., Culbertson, J. D., Cornwell, C. J., Mattick, L. R., Nygaard, M., In Press.
17. Makinen, K. K., Soderling, E., J. Food Sci. 1980, 45, 367.
18. Ulrich, R., in "The Biochemistry of Fruits and Their Products" Vol. 1, Hulme, A. C., Ed., Academic Press, London, 1970. p 97, 102.
19. Benk, E., Alkohol-Industrie 1978, 91, 88.
20. Whiting, G. C., J. Sci. Food Agric. 1958, 9, 244.
21. Fernandez-Flores, E., Kline, D. A., Johnson, A. R., J. Assoc. Off. Anal. Chem. 1970, 53, 17.
22. Ryan, J. J., Dupont, J. A., J. Agric. Food Chem. 1973, 21, 45.
23. Akhavan, I., Wrolstad, R. E., J. Food Sci. 1980, 45, 499.
24. Cash, J. N., Sistrunk, W. A., Stutte, C. A., J. Food Sci. 1977, 42, 543.
25. Johnson, T., Nagel, C. W., Am. J. Enol. Viticult. 1968, 27, 15.
26. Coppola, E. D., Conrad, E. C., Cotter, R., J. Assoc. Off. Anal. Chem. 1978, 61, 1490.

27. Fernandez-Flores, E., Kline, D. A., Johnson, A. R., Leber, B. L., J. Assoc. Off. Anal. Chem. 1970, 53, 1203.
28. Hood, L., Personal Communication. 1980. U.S. Customs Service, Department of the Treasury, Washington, D.C.
29. Blumenthal, A., Helbling, J., Mitt Gebiete Lebensm. Hyg. 1977, 68, 419.
30. Wucherpfenning, K., Flussiges Obst. 1976, 214.
31. Timberlake, C. F., Bridle, P., in "The Flavonoids", Vol. 1, Harborne, J. B., Mabry, T. J., Mabry, H., Eds., Academic Press London, 1975, p. 235.
32. Van Buren, J., in "The Biochemistry of Fruits and Their Products". Vol. 1, Hulme, A. C., Ed., Academic Press, London 1970, p. 278.
33. Wrolstad, R. E., "Color and Pigment Analyses in Fruit Products". Oregon Agricultural Experiment Station Bulletin 624, Corvallis, OR, 1976.
34. Mattick, L. R., Weirs, L. D., Robinson, W. B., J. Assoc. Off. Anal. Chem. 1967, 50, 299.
35. Fitelson, J., J. Assoc. Off. Anal. Chem. 1967, 50, 293.
36. Fitelson, J., J. Assoc. Off. Anal. Chem. 1968, 51, 937.
37. Barritt, B. H., Torre, L. C., J. Chromatogr. 1973, 75, 151.
38. Wrolstad, R. E., Struthers, B. J., J. Chromatogr. 1971, 55, 405.
39. Ryan, J. J., Dupont, J. A., J. Assoc. Off. Anal. Chem. 1973, 56, 743.
40. Withy, L. M., Heatherbell, D. A., Strachan, G., N.Z. J. Sci. 1978, 21, 91.
41. Culbertson, J. D., Wrolstad, R. E., in "Fruit and Berry Concentrate Processing in the Pacific Northwest" Third Quarterly Report to Pacific Northwest Regional Commission. Department of Food Science & Technology, Corvallis, OR., 1978; p. 1-5.
42. Sistrunk, W. A., Cash, J. N., J. Food Sci. 1973, 38, 807.
43. Li, K. C., Woodroof, J. G., J. Agric. Food Chem. 1968, 16, 534.
44. Ryan, J. J., J. Assoc. Off. Anal. Chem. 1972, 55, 1104.
45. Chang, Y. H., "Chemical Analysis of Acid Precipitated Colloid of Ripe Bartlett Pear Juice". M.S. Thesis. Oregon State University, Corvallis, OR 1979, p. 37.
46. Kliewer, W. M., Am. J. Enol. Viticul. 1968, 19, 166.
47. Abers, J. E., Wrolstad, R. E., J. Food Sci. 1979, 44, 75.
48. Boland, F. E., Blomquist, V. H., Estrin, B., J. Assoc. Off. Anal. Chem. 1968, 51, 1203.
49. Otteneder, H., Lebensmittelchemie und Gerichtliche Chemie 1978, 32, 56.

RECEIVED April 6, 1981.

8

The Role of Humidity, Temperature, and Atmospheric Composition in Maintaining Vegetable Quality During Storage

L. VAN DEN BERG

Division of Biological Sciences, National Research Council of Canada, Ottawa, Canada K1A 0R6

Rising energy costs favour local long term storage of vegetables over long distance transport from distant growing areas. The energy required to truck a ton of carrots from Florida or Texas, U.S.A., to Toronto, Canada, for example, is more than sufficient to store the carrots under refrigeration for one year. Local growers also benefit when storage of vegetables is profitable, a point not lost on politicians or local growers themselves. Hence, growers are encouraged to increase their storage capacity for vegetables.

The increased interest in storage of vegetables has led to a demand for up-to-date and valid information on optimum storage conditions, especially temperatures, relative humidity and atmospheric composition (C.A. storage). While there is widespread agreement on the optimum storage temperature for many vegetables, the same cannot be said for relative humidity. Recommended levels of relative humidity for the storage of fresh vegetables have traditionally been a trade-off between desiccation of the products by a low humidity on the one hand and increased decay by a high humidity on the other (1). The recommended optimum level of relative humidity has gradually edged closer to 100%, because it was found that decay was not quite as serious a problem at high humidity as it once was thought to be. Nevertheless, there is still a great reluctance on the part of horticulturists and storage operators alike to increase the recommended humidity level to the saturation point (100%). It is feared that such a high humidity would lead to a drastic increase in decay, either because of the humidity level itself or because condensation of water on the product becomes unavoidable in practice at this level.

The development of the jacketed storage (2-8) led to a practical and economical way of maintaining a saturated atmosphere in the load space and this put new urgency on the question of the optimum level of relative humidity for fresh produce. Also, use of other systems of direct humidifcation (9,10) and of plastic packaging (11) raised questions of optimum humidity levels.

0097-6156/81/0170-0095$05.00/0

Extensive storage tests with vegetables have therefore been made at the National Research Council Laboratories of Canada (6,12-21) and results have led to a broader and better understanding of the role of humidity in maintaining vegetable quality.

This paper presents unpublished and published information from studies in NRC's laboratories on the effect of relative humidity on decay and other quality factors during storage. Vegetables included were beets, Brussels sprouts, cabbage, carrots, cauliflower, celery, Chinese cabbage, leeks, parsnips and rutabagas. Tests always included a comparison between the "at" or near saturated level of relative humidity (98-100%) and one or more lower levels (85-95%). The length of storage involved varied from a few weeks for cauliflower up to one year for carrots. During this time frequent observations were made on weight loss, decay, appearance, color, firmness and crispness. Many of the tests also involved several storage temperatures (in the range 0 to 7°C) and modified atmospheres (0.5-10% oxygen, 1-11% carbon dioxide). Some basic studies on the effect of humidity on survival and growth of decay-causing molds and on vegetable resistance to pathogens were also made. The paper concludes with a summary on the effects of relative humidity, temperature and atmospheric composition on storage life of several vegetables.

Materials and Methods

Sample Material. Vegetables were obtained from major growing areas in the provinces of Prince Edward Island, New Brunswick, Quebec and Ontario in Canada. Carrots, celery, Chinese cabbage, leeks and parsnips were obtained from muck soils; beets, Brussels sprouts, cabbage, cauliflower and rutabagas from loamy clay soils. Cultivars selected were those generally stored in the growing areas. These vegetables were mature at harvest time, were handled carefully after harvest, and were placed under test conditions usually within 24 hours after harvest. They were not washed, cleaned or graded, except for discarding obviously damaged specimens and loose soil.

Storage Tests. Tests with about 35L samples were made in gas- and vapor-proof plywood cabinets (holding two 35L sample boxes each) in which temperature, relative humidity and atmospheric composition could be controlled accurately. The sterilized sample boxes were equilibrated at the storage relative humidity prior to the tests to avoid moisture absorption by the wood during tests (22). Modified atmospheres were obtained by flushing cabinets continuously with the desired gas mixture.

Larger scale tests, to determine the effect of sample size on decay at high relative humidity, were made in small completely jacketed rooms, used without internal forced air circulation. During tests, the relative humidity in the open space was 97-99%

and air temperature was controlled within ±0.3°C. Produce was generally stored in pallet boxes of a common commercial design (1.2 x 1.2 x 0.9 m, 2-8 boxes per test). Observations were made on decay, weightloss and quality during the storage period.

Basic Aspects. The fundamental studies focussed on the effect of relative humidity on host-parasite relationships during decay formation. Aspects studied were epidermis growth during storage, changes in fungistatic properties of carrots during storage, survival of pathogenic microorganisms during storage (23) and production and stability of pectolytic enzymes by these microorganisms (24).

Results and Discussion

Beets. Beets lost moisture readily at less than saturated humidity levels (Table I), resulting in soft spongy beets. Beets stored at 98-100% RH stayed solid and fresh. The optimum temperature for reducing decay was 4-5°C, rather than 0-1 or 2-3°C. However, sprouting increased with temperature but a sprout inhibitor could be used. Modified atmospheres at 0-1°C (about 1%O_2, 5-6%CO_2) caused rapid development of decay, especially mold. Damage during harvest, and stems and leaves not removed during harvest increased decay.

Brussels sprouts. Quality losses in Brussels sprouts caused by a less than saturated humidity level became apparent after about 6 weeks of storage. Brussels sprouts lost some of their green color, wilted and shrivelled, particularly when stored on the stalk. After about three months of storage, decay and trimming losses were substantially higher at the lower humidity (Table II).

Cabbage. Storage of cabbage at a relative humidity near or at saturation (98-100%) reduced decay (trimming loss), weight loss and color loss substantially in comparison with storage at 90-95% relative humidity at all temperatures studied (Table III). As a result of the reduced weight loss at the higher humidity (mostly less than 0.5% per 30 days as compared to about 2% per 30 days at 90-95%), cabbage remained firm and crisp during its storage life and retained its green color longer. Rooting and internal growth were not affected by relative humidity but depended markedly on temperature: storage life at 3.5-4.5°C and 7-8°C was limited to 4-5 months and 2-3 months respectively by internal growth and rooting; at 0-1°C these were not significant after 7 months. Low concentrations of oxygen and high concentrations of carbon dioxide further retarded the loss of green color of cabbage, and atmospheres containing 1-3% oxygen and 5% carbon dioxide caused an additional small reduction in decay. These results with modified atmospheres agree with findings in the literature (25,26). Oxygen levels below 1%, particularly at temperatures at or

Table I. Effect of relative humidity and temperature on decay and weight loss of Detroit Red beets after 25 weeks of storage (range includes results for 3 storage seasons).

RH, %	Temperature °C	Losses due to decay, % of initial weight	Rate of weight loss, %/30 days
85-95	0-1	40-100	1.5-4.0
98-100	0-1	25-35	0.1-0.5
98-100	2-3	10-20	0.1-0.5
98-100	4-5	0-10	0.1-0.5

Table II. Effect of relative humidity on decay and trimming loss and on weight loss of Brussels sprouts, cauliflower, celery and Chinese cabbage stored at 0-1°C for two or more seasons.

Vegetable and cultivars	RH, %	Storage time, weeks	Losses due to decay and trimming, % of initial weight	Rate of weight loss, %/30 days
Brussels sprouts (Jade Cross)	90-95 98-100	11-12 11-12	40-100 20-30	3.4-5.2 0.3-0.4
Cauliflower (Imperial 1006, Super Junior and Self-blanche)	90-95 98-100	5-6 5-6	15-30 20-30	3.0-3.9 0.6-1.4
Celery (Utah 15 and Utah 52-70)	90-95 98-100	11-12 11-12	70-75 45-65	2.2-3.1 0.4-0.5
Chinese cabbage (Michihli)	90-95 98-100	11-14 11-14	25-75 20-40	0.6-2.3 0-0.2

Table III. Effect of atmospheric composition and temperature on decay and trimming loss and on weight loss of green cabbage after 12 to 30 weeks of storage at 98-100% RH (averages for up to 5 storage seasons, "Evergreen" type cultivars).

Temperature, °C	Atmospheric composition, % O_2	CO_2	Losses due to decay and trimming, % of initial weight	Rate of weight loss, %/30 days
0-1	>20	<1(air)	20-30 (30 weeks)	0.3-0.7
0-1	10	11	40-70 (30 weeks)	0.2-0.5
0-1	6	5	30-50 (30 weeks)	0.2-0.5
0-1	1-3	5	10-30 (30 weeks)	0.2-0.5
3.5-4.5	>20	<1(air)	20-40 (20 weeks)	0.5-0.9
7-8	>20	<1(air)	10-30 (12 weeks)	1.2-1.6

slightly below 0°C sometimes caused internal breakdown. Results with red cabbage (Mammoth Red Rock) were similar to those obtained with green cabbage, and large scale tests gave results similar to those obtained in small scale tests. Studies in which water was condensed on the cabbage surfaces during storage showed that condensation was, if anything, beneficial rather than harmful in terms of quality and decay. This was also the case for carrots, celery and rutabagas.

Carrots. Decay was substantially less at the high level of relative humidity than it was at the lower level (90-95%), and depended on temperature (Table IV). The type of decay varied from relatively small (8 mm or less) dry brown lesions prevailing at 0-2°C to watery soft rot spot, prevailing at 3-8°C. Additional tests with maleic hydrazide treated carrots showed that this treatment did not significantly affect decay. Storage in modified atmospheres, on the other hand, increased decay markedly. Decay was not affected by sample size. In a few tests decay was slightly less at 2-5°C than at 0-1°C in agreement with findings by Smith (27).

Weight loss at 98-100% was only a fraction of that at 90-95 % (Table IV). Since a weight loss of 5-7% causes appreciable softening and wilting of carrots (28), these results indicate that most carrots would not be saleable if held more than four to six months at 90-95% relative humidity, even under ideal temperature conditions. Weight loss did not affect rooting and sprouting markedly although temperature did in carrots not treated with sprout inhibitor.

Cauliflower. The storage life of cauliflower was limited by brown spots appearing in 3-4 weeks on the curds, regardless of relative humidity. Low humidities caused yellowing and wilting of the wrapper leaves, but the storage life was too short for humidity to be a significant factor (Table II).

Celery. Celery stored at 98-100% relative humidity remained in marketable condition for at least two weeks longer than when held in a directly cooled storage at a lower humidity (3-3 1/2 months compared to 2 1/2-3 months). In both instances, mold growth determined storage life (Table II). The celery at the higher humidity was also appreciably more crisp and of better, greener color during the latter part of the storage period. An atmosphere of 3%O_2, 5%CO_2 caused the green color to stay longer and reduced decay to some extent. Loose packing of bunches in boxes also reduced decay.

Chinese cabbage. Chinese cabbage is also quite susceptible to moisture loss and associated quality loss (wilting, yellowing) (Table II). Trimming losses after about 3 months storage were greatly reduced at 98-100% as compared to a lower humidity. In contrast with other leafy vegetables, a modified atmosphere (1-3%O_2, 5%CO_2) did not reduce decay and color loss.

Leeks. Leeks are an exceptional vegetable in that they store well in a partially frozen condition at -1 to -1.5°C (Table V). Such a narrow range of temperatures is virtually impossible to maintain commercially, and a temperature close to 0°C is therefore optimal commercially. Decay and wilting are significantly reduced by high humidity, especially when combined with a low oxygen - high carbon dioxide atmosphere (1-3%O_2, 5%CO_2). Under the latter conditions, color is retained almost unchanged during the storage life.

Parsnips. Results obtained with Hollow Crown parsnips (Table IV) were similar to those obtained with carrots, in that decay at 98-100% relative humidity was less than or at most equal to that at 90-95%. At 0-1°C, the weight of decaying parsnips after 9 months of storage amounted to 10% or less of the pre-storage weight at both humidities, while at 3.5-4.5°C decay amounted to 10-20% at 98-100%, and 10-30% at 90-95% relative humidity. Rates of weight loss at these two humidities were 0.0-0.6 and 0.6-2.1% per 30 days, respectively, causing very marked differences in firmness.

Rutabagas. Decay in Laurentian rutabagas was independent of relative humidity and varied from less than 10% of prestorage weight after 9 months of storage at 0-1°C to 5-15% at 3.5-4.5°C at both humidities tested (90-95 and 98-100%). Decay took the form of soft rot spots which usually were readily trimmed off, and was not affected by the sample size. Rates of weight losses were 0.7-2.3 and upto 0.2% per 30 days respectively at these humidities. The higher rate of weight loss caused a noticeable softening after 9 months. Sprouting was not affected by relative humidity within the range studied.

Basic Aspects. The unexpected decrease in decay at 98-100% relative humidity as compared to 90-95% in many instances indicated a need to understand some of the underlying host-parasite relationships. With carrots as host, it was found that *Botrytis cinerea* was the most important pathogen, followed by *Sclerotinia sclerotiorum*.

Of the four aspects of host-parasite relationships studied (carrot periderm thickness, fungistatic properties of carrot tissue, survival of pathogens, and pectolytic enzyme production by the pathogens) only the enzyme production was affected by relative humidity in a way parallel to the effect of relative humidity on decay. It was found that *B. cinerea* produced substantially more enzymes on the surface of carrots stored at 90-95% than on those stored at 98-100%. Results also indicated that this higher enzyme production was caused by a lesser availability of nutrients

Table IV. Effect of relative humidity and temperature on decay and weight loss of carrots, parsnips and rutabagas stored for 35 weeks (average for 2 to 8 storage seasons).

Vegetable and cultivar	RH, %	Temperature, °C	Losses due to decay, % of initial weight	Rate of weight loss, %/30 days
Carrots (Nantes Chantenay and Goldpack)	90-95	0-1	20-40	1.1-1.6
	98-100	0-1	10-20	0.1-0.3
	98-100	2-3	10-20	0.3-0.5
	90-95	3.5-4.5	40-60	1.8-2.4
	98-100	3.4-4.5	10-25	0.4-0.8
Parsnips (Hollow Crown)	90-95	0-1	<10	0.6-1.0
	98-100	0-1	<10	0-0.4
	90-95	3.5-4.5	10-30	0.7-2.1
	98-100	3.5-4.5	10-20	0.2-0.6
Rutabagas (Laurentian)	90-95	0-1	<10	0.7-1.1
	98-100	0-1	<10	0-0.2
	90-95	3.5-4.5	5-15	1.9-2.3
	98-100	3.5-4.5	5-15	0-0.2

Table V. Effect of relative humidity, temperature and atmospheric composition on decay and trimming losses and weightloss of leeks after 14-18 weeks of storage (averages for 2-4 storage seasons; Unik and Elephant cultivars).

RH, %	Temperature, °C	Atmospheric composition, %		Losses due to decay and trimming, % of initial weight	Rate of weightloss %/30 days
		O_2	CO_2		
98-100	-1.0 to -1.5	>20	<1(air)	20-30 (14-18 weeks)	0-0.3
90-95	0-1	>20	<1(air)	45-75 (14 weeks)	2.4-2.6
98-100	0-1	>20	<1(air)	25-55 (14 weeks)	0.5-0.9
98-100	0-1	5	5	25-30 (18 weeks)	0-0.4
98-100	0-1	3	5	25-30 (18 weeks)	0-0.4
98-100	0-1	0.5-1	5	20-30 (18 weeks)	0-0.4

at the lower humidity level. It was noted that at 98-100% relative humidity, molds grew luxuriantly on the surface of many vegetables without causing decay or other permanent damaging effects. It was also found that the pectolytic enzymes were deactivated more rapidly on carrots stored at 98-100% than on carrots stored at 90-95%. Since these enzymes play an important role in the penetration of the pathogen into the host, the difference in enzyme production at different relative humidities is likely a major factor in the effect of relative humidity on decay.

Storage life of vegetables. The effects of relative humidity, temperature and atmospheric composition on the storage life of vegetables discussed earlier are summarized in Table VI. In this Table, storage life has been defined as the period during which the loss of produce caused by culling and trimming necessitated by weight loss (wilting and shrivelling), yellowing, rooting, sprouting and decay did not exceed 20-30%. This criterion was based on the following considerations:

i) Once losses reach this level, the rate of deterioration usually increases rapidly.

ii) For up to 20-30% trimming losses, vegetables such as cabbage, celery, Chinese cabbage, and leeks retain an identity closely related to the freshly harvested product.

iii) The labor and handling costs of trimming and discarding more than 20-30% of the produce is likely to be prohibitive.

The range of storage life given in each case includes variability due to cultivar, storage season and growing area. Factors responsible for limiting storage life are listed (in order of importance) in each instance. It should also be pointed out that the quality of produce stored at the higher level of humidity was usually better at the end of the indicated storage life than that stored at the lower humidity because of the greater crispness and firmness resulting from reduced moisture loss. As Table VI shows, 98-100% RH resulted in a longer storage life than 90-95% RH for most vegetables. The storage life given for 90-95% RH are in good agreement, for the most part, with data given by Lutz and Hardenburg (1) for comparable conditions.

Conclusions and Summary

Results of extensive small and large scale laboratory tests have shown that:

1. Decay of most vegetables during refrigerated storage at 98-100% relative humidity was less than or about equal to that during storage at lower humidity levels.

2. Weight losses during storage at 98-100% relative humidity were substantially less than during storage at lower humidities, particularly for those vegetables with a high coefficient of transpiration (29).

Table VI. Effect of relative humidity and temperature on estimated storage life of vegetables stored in air.

Vegetables	Cultivars tested	Storage temp., °C	RH, %	Storage life,[a] weeks	Main factors limiting storage life
Beets	Detroit Red	0 to 1	90-95	20-25	Wilting,[b,c] decay
			98-100	20-30	Decay
		2 to 3	98-100	25-35	Sprouting, decay
		4 to 5	98-100	25->35	Sprouting
Brussels sprouts[e]	Jade Cross	0 to 1	90-95	6-8	Wilting, decay, yellowing
			98-100	10-12	Decay
Cabbage[e]	Penn State	0 to 1	90-95	14-22	Decay, wilting, yellowing
	Ballhead		98-100	18-26	Decay, yellowing
	Storage Green	3.5 to 4.5	90-95	10-14	Decay, wilting, yellowing
	Houston Evergreen		98-100	12-16	Decay, internal growth, rooting
		7 to 8	90-95	8-12	Internal growth, rooting
			98-100	8-12	Internal growth, rooting
Carrots	Special Long	0 to 1	90-95	15-30	Decay and softening
	Type Nantes		98-100	30-40	Decay, rooting and sprouting[d]
	Royal Dutch Nantes	2 to 3	98-100	30-40	Decay, rooting and sprouting
	Royal Chantenay	3.5 to 4.5	90-95	15-30	Decay and softening
	Gold Pack		98-100	20-40	Decay, rooting and sprouting
		7 to 8	98-100	20-40	Decay, rooting and sprouting
Cauliflower	Imperial 1006	0 to 1	90-95	3-4	Brown spots on heads
	Super Junior		98-100	3-4	Brown spots on heads
	Self-blanche				
Celery[e]	Utah 15	0 to 1	90-95	8-10	Wilting, yellowing, decay
	Utah 52-70		98-100	10-12	Decay, yellowing
	Merveilleux	0 to 1	98-100	4-5	Decay

Chinese cabbage	Michihli	0 to 1	90-95	8-12	Wilting, yellowing, decay
			98-100	10-14	Decay
Leeks[e]	Unik	-1 to -1.3	98-100	14-18	Decay
	Elephant	0 to 1	90-95	7-10	Decay and wilting
			98-100	9-14	Decay
Parsnips	Hallow Crown (maleic hydrazide treated)	0 to 1	90-95	20-30	Shrivelling, decay
			98-100	>35	Decay
		3.5 to 4.5	90-95	15-25	Shrivelling, decay
			98-100	≥35	Decay
Rutabagas	Laurentian	0 to 1	90-95	≥35	Softening, decay, sprouting
			98-100	≥35	Decay, sprouting
		3.5 to 4.5	90-95	≥35	Softening, decay, sprouting
			98-100	≥35	Sprouting, decay

[a] A loss of 20-30% of prestorage wt as a result of trimming and culling necessitated by yellowing, shrivelling, wilting, softening and decay alone or in combination, was assumed to end the useful storage period.

[b] Weight losses generally amounted to 0 to 0.5%/30 days at 98-100% RH and 0.7-2.0%/30 days at the lower humidity levels.

[c] A total moisture loss of 5-6% causes the first noticeable softening or wilting of root crops and solid cabbage heads, while a total moisture loss of over 8% makes them virtually unsaleable. With leafy vegetables, a weight loss of 3 to 4% becomes noticeable (outside leaves wilted) and requires trimming.

[d] Sprouting depended on the use of sprout inhibitor.

[e] Modified atmospheres (<3% O_2 + 5% CO_2) resulted in longer storage life (about 2 weeks for celery, 2-6 weeks for cabbage and leeks).

3. The reduced weight loss at the high humidity results in firmer, crisper and higher quality vegetables, and, in the case of cabbage and celery, substantially greener color.

4. Optimum storage conditions for these vegetables therefore include a relative humidity of 98-100%, as close to saturation as possible, such as can be obtained in jacketed rooms.

5. The reduced decay at 98-100% relative humidity as compared to that at 90-95% for carrots and possibly several other vegetables is at least in part caused by a lower production of pectolytic enzymes by the pathogenic micro-organisms at the higher humidity level.

Literature Cited

1. Lutz, J.M.; Hardenburg, R.E. "Agriculture Handbook 66", U.S.D.A., Washington, D.C., U.S.A. 1968.
2. Jorgensen, E.G. Can. Refrig. and Air Cond. 1968, 34(7), 16.
3. Jorgensen, E.G. ASHRAE Bulletin 1973, CH-73-7, 35.
4. Lentz, C.P. Proc. 10th Int. Congr. Refrig. 1959, 3, 298.
5. Lentz, C.P.; Phillips, W.R. Proc. 10th Int. Congr. Refrig. 1959, 3, 309.
6. Lentz, C.P.; Rooke, E.A. Food Technol. 1957, 11, 257.
7. Lentz, C.P.; Jorgensen, E.G.; Sawler, R. Can. Inst. Food Technol. J. 1971, 4, 19.
8. Phillips, W.R.; Lentz, C.P.; Rooke, E.A.; Rutherford, W.M. Can. Refrig. and Air Cond. 1961, 27(12), 20.
9. Krahn, R. Acta Hort. 1974, 38, 443.
10. Meredith, D. ASHRAE Bulletin 1973, CH-73-7, 29.
11. Hardenburg, R.E. ASHRAE Bulletin 1973, CH-73-7, 19
12. Van den Berg, L.; Lentz, C.P. Food Technol. 1966, 20, 104.
13. Van den Berg, L.; Lentz, C.P. Intern. Institute of Refrig. Bull. Annex. 1966, 1, 335.
14. Van den Berg, L.; Lentz, C.P. Food in Canada 1966, 26(7), 16.
15. Van den Berg, L.; Lentz, C.P. J. Amer. Soc. Hort. Sci. 1972, 97, 431.
16. Van den Berg, L.; Lentz, C.P. J. Food Sci. 1973, 38, 81.
17. Van den Berg, L.; Lentz, C.P. J. Am. Soc. Hort. Sci. 1973, 98, 129.
18. Van den Berg, L.; Lentz, C.P. ASHRAE Bulletin 1973, CH-73-7, 12.
19. Van den Berg, L.; Lentz, C.P. Can. Inst. Food Sci. Technol. J. 1974, 7, 260.
20. Van den Berg, L.; Lentz, C.P. Acta Hort. 1977, 62, 197.
21. Van den Berg, L.; Lentz, C.P. Acta Hort. 1977, 62, 209.
22. Brown, W.G.; Lentz, C.P. Can. J. Technol. 1956, 34, 366.
23. Van den Berg, L.; Lentz, C.P. Can. J. Bot. 1968, 46, 1477.
24. Van den Berg, L.; Yang, S.M. Can. J. Bot. 1969, 47, 1007.
25. Isenberg, F.M.R. Hort. Reviews 1979, 1, 337.

26. Isenberg, F.M.R.; Sayles, R.M. J. Am. Soc. Hort. Sci. 1969, 94, 447.
27. Smith, W.H. J. Food Technol. 1967, 2, 89.
28. Lentz, C.P. Food Technol. 1966, 20(4), 201.
29. Van den Berg, L.; Lentz, C.P. Can. Inst. Food Technol. J. 1971, 4, 143.

RECEIVED April 3, 1981.

9

The Improvement of Flavor in a Program of Carrot Genetics and Breeding

P. W. SIMON and C. E. PETERSON
U.S. Department of Agriculture, Science and Education Administration, Agricultural Research, Department of Horticulture, University of Wisconsin, Madison WI 53706

R. C. LINDSAY
Department of Food Science, University of Wisconsin, Madison, WI 53706

The genetic improvement of vegetable quality is a complex task which requires the concerted efforts of several disciplines. The ultimate needs to be satisfied are those of the consuming public. Therefore, long-range plans for improvement must include a determination of positive and negative quality attributes of relevance to nutritional needs and sensory preferences of consumers. Horticultural characteristics for large-scale production and marketing must also be maintained while quality improvements are being made. When quality attributes to be altered have been determined and means of simultaneously fulfilling production criteria established, improvement can proceed.

For reasonable genetic gains in quality to be made, attributes under consideration should be readily measurable using simple techniques, they should be stable or at least predictable over the usual range of growth, storage, and processing conditions, and inheritance patterns of attributes should be known. Finally, methods for incorporating the improved attributes into the existing genetic, physiological and cultural system must be determined.

Unfortunately, the environmental and genetic variability observed with many quality factors is either complex or unknown. Furthermore, the biological components of quality are difficult to assess (e.g. flavor, protein efficiency ratio), so that consumer panels or other complex biological assays must be used to measure quality.

Carrots (*Daucus carota* L.) are of considerable nutritional significance because they are the most important vegetable source of provitamin A in the U.S. diet, providing 14% of the total (*17*). Varietal and seasonal differences in carrot flavor have been reported since the introduction of carrots to Europe from the Mideast in the 12th century (*3*). This report details some of the particular problems and potential solutions for the genetic improvement of carrot quality.

0097-6156/81/0170-0109$05.00/0

Perspective of Quality in U.S. Carrot Production and Improvement

Carrot Production. The ability to improve the organoleptic and nutritional quality of carrots by changing production and distribution requirements is best understood by considering some statistics and regulations of production. The majority of carrots grown in the U.S. are for the fresh market which accounts for over 80% of the per capita consumption. The remainder is for processed carrots (25). Marketable yield for fresh market carrots is a function of root size, shape, and external appearance. Eating quality is not a factor in grading fresh vegetables for sale, even though it may be very important for repeat sales and for establishing consumer preferences.

Carrot Genetics and Breeding. The ability to introduce quality improvement as a major objective in carrot breeding is easier to understand with an overview of the numbers and operations of such programs. Three federal or state programs and several private seed firms currently perform all of the breeding and genetic improvement for carrots in the U.S. Therefore, no single research program can confine all of its efforts to improving culinary quality. The trend for new cultivars is toward hybrid carrots, which provide a more uniform, predictable crop for yield and quality than does the older method of open-pollinated varietal improvement.

The U.S.D.A. carrot improvement program currently includes approximately 50 inbred carrot lines, and annually tests several hundred hybrid crosses and numerous segregating populations. In addition to culinary and nutritive quality parameters, characteristics being improved include root color, shape, and size, disease resistance, root storage ability, uniformity of flowering, and seed production. All of these characteristics must be maintained at an acceptable level while attempting to improve quality.

One feature of the carrot life cycle which must be considered is its biennial habit. Carrot roots must be refrigerated for up to 8 weeks after harvest before they will flower and consequently set seed, and only the crown third of the root is needed for seed production. This allows laboratory and sensory analysis to be made on the remaining portion of the root while the cold treatment is underway, and only selected roots are used to produce seed.

Carrot Quality Improvements. The lack of definite eating quality standards, and the difficulty of quickly and easily assessing carrot quality makes the grading of culinary quality in carrots as they come from production areas to the consumer unfeasible. This places the burden of quality improvement on altered production practices, varieties with genetically improved quality over a wide range of environments, or both.

Since acceptance of new cultivars depends on a favorable response from growers, short-term breeding and genetic goals tend to be producer-directed. This is satisfactory for most consumer needs, but as stated above, culinary and nutritive traits can be overlooked. Long-range improvement programs should also direct research efforts to consumer needs. GRAS (generally regarded as safe) regulations for nutrients and toxicants set by the Food and Drug Administration further emphasize consumer welfare (17).

Thus, the improvement of carrot quality can be a major objective if the breeder is willing to pursue long-term research goals while concurrently placing adequate emphasis on marketable yield. With such a program, the steps for quality improvements are to define important quality attributes, ascertain the feasibility of their improvement, and establish techniques for screening the large number of samples that must be tested.

Steps in the Improvement of Carrot Quality

Choosing Quality Attributes to be Tested. The choice of nutritional and toxic metabolites to be considered and methods of analysis are largely dictated by health scientists. As a result, carotenoid (9), myristicin and falcarinol (30) levels have received some attention in carrots. Choosing sensory characteristics is more difficult, and requires an accurate assessment of influential attributes along with identification of suitable terms for communication. In the case of carrots, undesirable and variable flavors are commonly mentioned culinary quality attributes needing improvement. Bitter flavor, often developing in stored carrots, results from the synthesis of "isocoumarins" in the presence of ethylene (6, 22). More recently, the need for sweeter and less harsh ("turpentiney", "oily") flavor has been recognized by carrot growers as desirable, and formal sensory evaluations have demonstrated that panelists can distinguish between extremes for these attributes (18, 20).

Determining Means for Implementing Quality Improvement. Culinary quality improvement cannot be attributed exclusively to selection of preferred genotypes. Altered cultural practices and controlled atmospheric storage have resulted in flavor improvement in many fruits and vegetables, including bitterness in carrots, without creating new genotypes (10). Thus, it must be determined how strongly genetic and non-genetic sources of variation contribute to improved culinary quality. This determination should be made over the range of environments prevalent in important growing areas, storage conditions, and genetic stocks available so that the magnitude of potential genetic change can be estimated.

For carrots, genetic variation for sensory attributes is substantially greater than environmental (climate and soil) variation (18, 20, 21). This indicates an opportunity for substantial improvements of culinary quality in a program of carrot genetics and breeding.

Screening Techniques for Quality Factors in a Program of Genetics and Breeding. Important quality factors exhibiting genetic variation should be screened in every generation of the breeding cycle. The development of genetic stocks often requires ten to twenty years. Less important factors may only be screened at the beginning of the cycle, and again when the material is about to be released for public use. Simply inherited factors will need to be less frequently screened if the genetic purity of the original material is carefully controlled.

Components of nutritional quality are generally screened by accepted laboratory methods. Often modifications or shortcuts designed for the specific item to be samples are used to speed up such analyses [e.g. extremes in color of carrot closely estimate extremes in carotenoid levels (9)].

The ideal system for scoring culinary quality samples is with a trained, descriptive taste panel since this approximates the test made by the consumer. However, limitations of time and resources prohibit this approach with genetically segregating materials of a breeding program. For example, if the ten sweetest roots were sought in four different genetic or breeding stocks of 100 roots each, 400 roots would need to be scored by the panel. Since a typical descriptive sensory evaluation panel is comprised of 25 participants, and scores at most five samples at one sitting, the small available sample size and large number of roots to be scored would both rule out the use of panels for such preliminary screening. Sensory analysis is a well-established tool, however, for two other phases of this task: correlation of sensory and objective parameters (trained panel) and evaluation of the final product (consumer panel).

The ease of performing physical and chemical laboratory analyses in comparison to sensory evaluations has led to the use of selected laboratory (objective) data to predict sensory response. High correlations between selected objective data and sensory parameters are necessary, but adequate relationships have been determined in genetic variants of several fruits and vegetables, including grapes (15, 28), apples (27, 29), pears (26), raspberries (12), tomatoes (24), snapbeans (23), onions (16), and carrots (20). In addition to the relative ease of laboratory analyses, they offer greater speed and repeatability than sensory evaluation and they require much smaller samples. This latter feature is crucial where a sample usually consists of only one plant's root, fruit or leaves. Harsh flavor in carrots is related to levels of volatile terpene compounds, and a quantitative laboratory method for these compounds employing porous polymer trapping can be used to analyze up to 30 small (25 g) samples per day (19, 20). The use of an efficient volatile terpenoid analytical method also provides chemical information for determining the biogenetic pathways of these compounds. This information may suggest the most effective way to genetically fix these terpenoids at a desired level.

Sensory/objective parameter correlations are of particular interest to flavor chemistry research where reconstitution or enhancement of an organoleptic response by a correlated compound or group of compounds provides support for correlations noted. For example, flavor notes of apple (8), blueberry (11), tomato (24), and orange juice (1, 2) have been reconstituted. Sweetness in carrots is enhanced by added fructose with no significant diminution of harsh flavor (18) whereas harsh flavor has been elicited in mild carrots by the addition of suitable levels of the major volatile terpenoids of carrot (Table I).

In this study a mixture of terpinolene, caryophyllene, γ-terpinene, limonene, and myrcene (70:14:9:5:2) was dissolved in an equal amount of propylene glycol, blended for 30 sec. at high speed in water to give a concentration of 1% terpenoids, and applied to carrot slices. The terpenoid mixture was administered as a 25 μl droplet between two 2.5 g carrot slices to raise the total terpenoid level of the mild inbred B9304 (10 ppm) by 50 ppm which was nearly that of the harsh inbred B0493 (68 ppm). Introduction of the terpene solution between carrot slices proved to be a successful means of providing a uniform flavoring release and mixing during assessment of the samples by panelists. The added terpene compound mixture yielded a harsh flavor intensity in the mild B9034 carrots similar to that of the harsh B0493 carrots. While not specifically evaluated, the harsh flavor quality of the naturally-flavored B0493 carrots was very similar to that of the artificially-flavored B9304 carrots. Perceived sweetness appeared to be suppressed by the higher levels of terpene compounds although the difference was not statistically significant. However, both harsh-flavored carrot samples were significantly less preferred than the mild-flavored B9304 samples indicating the important influence of volatile terpenoid concentrations on the culinary quality of carrots.

With the chemical analysis of 30 or more samples per day in the laboratory, it is possible to secure enough genetic data to assess the feasibility of improving quality parameters, to determine biosynthetic pathways of compounds important to quality, and to correlate objective data with sensory parameters. Furthermore, these methods can be used to screen genetic materials in the numbers needed in breeding for quality. It is not possible, however, to screen all stocks in such a program and still perform the operations necessary in a traditional program for improving marketable yield. A faster sampling technique must be used to accomplish this.

Several techniques are available to vegetable breeders for estimating quality factors rapidly, including refractive index for total sugar levels, specific gravity for dry matter content, visible color for pigment levels, dichloroindophenol for vitamin C levels (14), 2,4-dinitrophenylhydrazine for pungent compounds in onions (16), and IR reflectance for protein levels. Except total sugar measurements and pungent sulfur-containing compounds,

Table I. Mean Scores for Descriptive Sensory Analysis of Mild and Harsh Carrots with Added Terpene Flavor Mixture

Raw Carrot Samples	Sample Attributes		
	Intensity of Harshness[a]	Intensity of Sweetness[a]	Overall Preference[b]
	(------------Mean scores[c]	------------)	
Mild Inbred (B9304) + 1% Propylene Glycol	2.24 a	4.15 a	4.77 a
Mild Inbred (B9304) + Basic Terpene Mixture in 1% Propylene Glycol	4.33 b	3.67 a	3.19 b
Harsh Inbred (B0493) + 1% Propylene Glycol	4.68 b	3.60 a	3.16 b

[a]Scale: 1 = very weak; 7 = strong.

[b]Scale: 1 = dislike very much; 7 = like very much.

[c]a, b: Mean scores in same column followed by the same letter are not significantly different at 5% level; n = 30.

high speed laboratory techniques are not available for most flavor compounds. This again leads to the use of sensory analysis for organoleptic factors, but with a different approach.

Trained testers and expert judges have long been used in the quality control of many foods, including dairy products, beer, wine, brandy, coffee, and soy sauce. For carrots, trained taste panels noted a difference between genetic stocks of over 1.9 units on a 7 point scale for harsh flavor, and over 1.0 units for sweetness and preference with samples from three locations (18). With this great a difference and a standard error of the mean at around 0.20, only 2 or 3 assessments need to be made to discriminate between such genetic stocks with a 95% level of confidence (4). Thus, there is a good opportunity for using 2 or 3 evaluators, or only one with multiple assessments, to score for extremes in flavor between carrot samples. Since each evaluation needs to be made with only 3-5 g slices, 3 determinations can be made from the middle one-third of even a small (30 g) root without destroying the crown portion needed for seed production.

The effectiveness of this screening method is being tested with carrot populations segregating for harsh flavor. One evaluator (P.W.S.) has scored roots in triplicate, and putative uniformly extreme populations will be tested by a standard sensory evaluation panel after several cycles of selection. When selection is performed for only one attribute, several samples can be tested per minute. Less fatigue is experienced if samples are not swallowed and carryover influences from harsher samples can be greatly reduced with a rinse of carbonated water. Periodic checks of sensitivity are made by testing slices of known mild samples. Typically, an acceptable repeatability (±15%) is realized for over 80% of the samples tested. Those samples not falling into this range are not included in further testing. With this system of quality screening, 400 carrot roots can be scored for one flavor attribute per day.

Current Status and Future Prospects for Improvement of Carrot Culinary Quality

Steps to improve the culinary quality of carrots include the reduction of bitterness by altered postharvest storage techniques and the reduction of harsh flavor by genetic selection. The 25-fold genetic variation in "isocoumarin" levels also suggests a potential for genetic improvement of bitterness (5, 6). Dominance for mild flavor over harsh flavor in an F_1 hybrid grown in nine environments (21) and in a range of F_1 hybrids grown in two different environments (Table II) suggest excellent prospects for improving the flavor of hybrid carrots by including at least one mild flavored parent in the hybrid. An investigation of the inheritance of sugar types and levels may also provide direction for the genetic improvement of carrot flavor (7).

Table II. Harsh Flavor Intensity of Several Inbred and F_1 Hybrid Carrots Grown in Two Locations.[a]

Parent[b]	B10138	B9304	B6274	B524	B4367	B0493	B3615
	(Mean scores[c])						
B10138	1.2/1.0	1.2/1.0	1.5/1.0	2.0/1.0	2.1/2.3	2.5/1.5	2.2/2.6
B9304		1.2/1.0	1.3/1.0	2.2/1.2	2.3/2.0	2.9/3.0	2.4/2.0
B6274			2.1/1.5	2.1/1.4	2.1/1.5	2.9/1.5	4.7/4.8
B524				3.5/2.5	3.5/2.6	5.0/5.5	5.4/4.1
B4367					5.8/5.6	5.7/5.6	5.9/5.8
B0493						6.8/6.6	6.8/6.8
B3615							7.0/7.0

[a]Roots were grown at Zellwin Farms, Zellwood, FL (muck soil) and the Imperial Valley Field Station, El Centro, CA (sandy soil) during the winter (Oct., 1979-Feb. 1980).

[b]Scores for inbreds are on the diagonal.

[c]Scale: 1 - low intensity of harsh flavor; 7 = high intensity of harsh flavor; X/Y - FL score/CA score; n = 6 or 7.

Future activities include the improvement of techniques for selecting quality attributes already identified, and to add other attributes. A faster laboratory method for screening the "iso-coumarins", volatile terpenoids, sugar types, and carotenoids will be valuable. Other flavor attributes of both fresh and processed carrots need consideration. The potential for genetic improvement of carrot fiber, which is a high quality nutrient (13) that may influence texture, also warrants investigation.

With a better knowledge of the inheritance of quality factors and with improved methods for screening them, the improvement of culinary and nutritive components in a program of carrot breeding and genetics is feasible. A proper perspective of quality and marketable yield must be maintained, but quality characteristics should be included in genetic improvement programs.

Abstract

The improvement of raw carrot flavor requires a determination of desirable and objectionable sensory attributes, an efficient means to assess these attributes or factors related to them, and an understanding of how flavor can be altered genetically and environmentally. Using a wide range of genetic stocks grown in several environments, sensory evaluation indicated substantial variation in carrot flavor due to environmental and, to a greater extent, genetic factors. Statistical analyses suggested that volatile terpenes and sugars could account for the observed sensory variation, and reconstitution experiments with authentic compounds supported these observations. Genetic experiments indicated dominance for both mild flavor and low volatile levels, to make the use of mild-flavored inbred lines advisable in producing hybrid carrots. Methods for screening large, segregating carrot populations for flavor improvement are available and readily applicable in a carrot breeding program.

Literature Cited

1. Ahmed, E. M.; Dennison, R. A.; Dougherty, R. H.; Shaw, P. E. J. Agric. Food Chem. 1978a, 26, 192.
2. Ahmed, E. M.; Dennison, R.A.; Shaw, P. E. J. Agric. Food Chem. 1978b, 26, 368.
3. Banga, O. Euphytica 1957, 6, 54.
4. Basker, D. Chem. Senses Flavor 1977, 2, 493.
5. Carlson, B. C. Ph.D. Thesis, Michigan State University, MI, 1960.
6. Carlson, B.C.;Peterson, C.E.; Tolbert, N. E. Plant Physiol. 1961, 36, 550.
7. Freeman, R. E.; Simon, P. W.; Peterson, C. E. HortScience 1980, 15, 421.

8. Guadagni, D.C. Am. Soc. Test. Mater., Spec. Tech. Publ. 1968, 440, 36.
9. Laferriere, L.; Gabelman, W. H. J. Am. Soc. Hort. Sci. 1968, 93, 408.
10. Nilsson, T. Ed., Quality of Vegetables, Acta Hort. 1979, 93.
11. Parliment, T.H.; Kolor, M. C. J. Food Sci. 1975, 40, 762.
12. Pyysalo, T. Z. Lebensm. Unters.-Forsch. 1976, 162, 263.
13. Robertson, J. A.; Eastwood, M. A.; Yeoman, M. M. J. Sci. Food Agric. 1979, 30, 388.
14. Rymal, K. S.; Rice, C. A. HortScience 1976, 11, 23.
15. Schreier, P. C. R. C. Crit. Rev. Food Sci. Nutr. 1979, 12, 59.
16. Schwimmer, S.; Guadagni, D. G. J. Food Sci. 1962, 27, 94.
17. Senti, F. R.; Rizek, R. L. HortScience 1975, 10, 243.
18. Simon, P. W.; Peterson, C. E.; Lindsay, R. C. J. Am. Soc. Hort. Sci. 1980a, 105, 416.
19. Simon, P. W.; Lindsay, R. C.; Peterson, C. E. J. Agric. Food Chem. 1980b, 28, 549.
20. Simon, P. W.; Peterson, C. E.; Lindsay, R. C. J. Agric. Food Chem. 1980c, 28, 559.
21. Simon, P. W.; Peterson, C. E. HortScience 1980d, 15, 421.
22. Sondheimer, E. J. Am. Chem. Soc. 1957, 79, 5036.
23. Stevens, M. A.; Frazier, W. A. Proc. Am. Soc. Hort. Sci. 1967, 91, 274.
24. Stevens, M. A.; Kader, A. A.; Albright-Holton, M.; Algazi, M. J. Am. Soc. Hort. Sci. 1977, 102, 680.
25. U.S.D.A. Agricultural Statistics, U.S. Gov't. Printing Office (1979).
26. Visser, T.; Schaap, A. A.; DeVries, D. P. Euphytica 1968, 17, 153.
27. Visser, T.; Verhaeg, J. J. Euphytica 1978, 27, 753.
28. Wagner, R.; Dirninger, N.; Fuchs, V.; Bronner, A. Int. Symp. Quality Vintage 1977, 137, S. African Minist. Agric.
29. Williams, A. A.; Lea, A. G. H.; Timberlake, C. F. Am. Chem. Soc. Symp. 1977, 51, 71.
30. Wulf, L. W.; Nagel, C. W.; Branen, A. L. J. Agric. Food Chem. 1978, 26, 1390.

RECEIVED April 3, 1981.

10

Non-Bitter Hop Contributions to Beer Flavor

VAL PEACOCK and MAX DEINZER

Department of Agricultural Chemistry, Oregon State University, Corvallis, OR 97331

The main purpose of using hops in beer is to add bitterness to the final product. A series of compounds referred to as α and β-acids are responsible for this taste (1). A secondary, ill defined flavor (flavor referring to smell and taste) is also imparted to beer by brewing with "aroma hops." Not all hop varieties are considered "aroma hops," and there is evidence that the flavors imparted to beer by different aroma hops are different (2). There has been considerable controversy in recent years as to the nature and source of this flavor. Researchers have credited terpene alcohols (2, 3), humulene oxidation products (4, 5), multicyclic terpenoid ethers (6) and carotenoids (6) as being in part responsible for this flavor.

Floral Hop Aroma/Taste

The most easily definable hop contribution to beer aroma is a floral flavor note that certain hop varieties (not necessarily the traditional "aroma hop" varieties) impart to beer (2). Indications are (Table I) that the floral compounds linalool and geraniol are responsible for this aroma note. Geranyl isobutyrate, though present in the more floral beers, is probably in too low concentration to have a major effect on beer flavor. α-Terpineol is eliminated from consideration for the same reason. Linalool has been reported in beer at an estimated concentration of 34ppb (3) by Lindsay and at a concentration of 470ppb (7) by Tressl.

As the current techniques for measuring these trace organics in beer are somewhat dubious, and considering the difficulties different investigators have in agreeing on what the sensory thresholds are for these compounds in beer (8, 9), it is hard to say what relative importance to beer hop aroma/taste they have. Clearly linalool must be an important contributor to this aroma/taste even considering Meilgaard's threshold of 80ppb (8) for this compound in beer. Geraniol seems to be important only in beers brewed with certain hop varieties, whereas linalool is common to all the beers and hops investigated. Geraniol in hops

0097-6156/81/0170-0119$05.00/0

Table I. Hop Oil Components Found in Beer

Compound	Cascade[a] Beer	European[b] Mixture Beer	Cluster[c] Beer	Hallertauer[d] Beer	Threshold in Beer
Trans-Linalool oxide[e]	--	--	--	25ppb	--
Linalool[e]	200ppb	200ppb	200ppb	200ppb	27ppb[b]
α-Terpineol[e]	75	--	75	175	2000[f]
Geranyl Isobutyrate[g]	150	--	25	--	450[g]
Geraniol[g]	200	--	75	--	36[g]
4,4-Dimethylcrotonolactone[e]	50	100	200	2000	>40,000[e]
Humulene Epoxide I[e]	--	100	50	--	10 in water[h]
Caryolan-1-ol[e]	--	25	50	--	--
Nerolidol[e]	--	75	--	--	--
T-Cadinol[e]	100	200	50	150	--
δ-Cadinol[e]	--	100	--	--	--
Humulenol II[e]	250	250	--	500	500[e,i]
β-Eudesmol[e]	--	100	--	150	--
Humulol[e]	--	--	75	200	--

a. Commercial beer brewed with 60% Cascade and 40% Cluster hops. b. Commercial beer brewed with a mixture of European hops. c. A pilot brew made by a commercial brewery using Cluster hops only. d. Same as c., only using Hallertauer hops. e. Ref. No. 4. f. Ref. No. 8. g. Ref. No. 2. h. Ref. No. 6. i. Subsequent analyses showed a threshold of 2500 ppb.

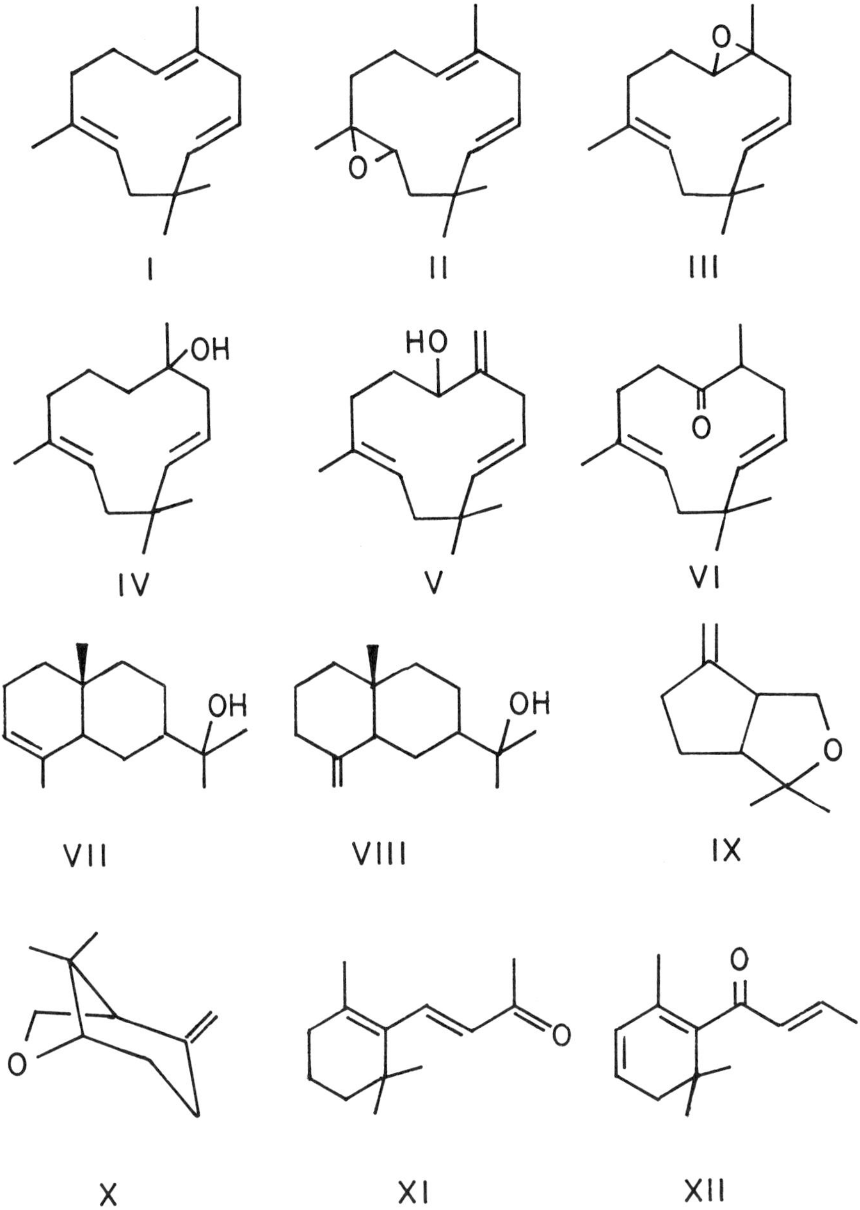

Figure 1. Structures of hop oil components. Key: I, humulene; II, humulene epoxide I; III, humulene epoxide II; IV, humulol; V, humulenol II; VI, humuladienone; VII, α-eudesmol; VIII, β-eudesmol; IX, hop ether; X, karahana ether; XI, β-ionone; and XII, β-damascenone.

is varietal specific. Geraniol is believed to be responsible for the much more intense floral aroma of beers brewed with Cascade hops (2). There is considerable disagreement as to what the sensory threshold of this compound is in beer, but both major investigators (2, 9) report that there is an unusually large variation of individual thresholds. Many persons (one-third) can perceive geraniol confidently at 25ppb and others cannot detect geraniol at even a few hundred ppb. To most people geraniol is apparently an important aroma/taste contributor in beer.

Geranyl isobutyrate is an important floral hop component as it is believed that much of the compound present in hops is hydrolyzed to geraniol, which is much more flavor active, during brewing (2). Transesterification of hop oil fatty-acid methyl esters to methanol and their corresponding ethyl esters is believed to take place during fermentation (6, 10). Hydrolysis of this ester during fermentation, therefore, would not be surprising. The ratio of geraniol to geranyl isobutyrate in Cascade-hopped and Cluster-hopped beers (Table I) as compared to their hop oils (Table II) tends to support this idea. It has been reported that fermentation of a 6% glucose solution with 50ppm geranyl isobutyrate added resulted in the hydrolysis of 15% of the ester (2). Hop storage, wort boiling and beer aging may result in further hydrolysis. As a result, it is the overall geraniol-geranyl isobutyrate content of the hop that should be considered as well as the linalool content. An attempt has been made to correlate the amount of this floral flavor one might expect from a beer brewed to the same bitterness level using different hop varieties. The yield of oil and bittering compounds from the hop was considered in this correlation. Cascade, Shin-shu-wase and Backa were reported as the varieties from which one would expect the most floral flavor and Perle and Hallertauer were reported to be low in floral flavor potential (2).

"Kettle Hop" or "Noble Hop" Aroma/Taste

"Kettle hop" aroma is an illusive flavor note reminding one of hops, imparted to beer by the vigorous boiling of "aroma hops" in the wort for up to a few hours. Brewers and hop flavor researchers currently do not agree on the chemical or sensorial nature of this flavor note. It is commonly described as spicy or herbal. There are no standard compounds used to illustrate this flavor note to flavor panelists.

This long wort boiling time is important in that most of the mass (80-90%) of typical hop oil is made up of terpene and sesquiterpene hydrocarbons which are either steam distilled out of the wort, polymerized or oxidized to more water soluble compounds during the process (4, 5, 11, 12). As a result, these hydrocarbons are not found in beer (4, 5, 7) and therefore are not responsible for this flavor. Investigators in this field do agree that this

Table II. Beer Floral Aroma/Taste Components in Hops[a,b]

Hop Variety	Linalool	Geraniol	Geranyl Isobutyrate
Cascade[c]	.85	.27	1.56
Cluster[c]	.44	.24	.60
Tettnanger[c]	.75	.03	.12
Hersbrucker[c]	.50	n.d.	n.d.
Hallertauer[c]	.41	n.d.	n.d.
Perle[c]	.26	n.d.	n.d.
Northern Brewer[c]	.28	.06	.12
Shin-shu-wase	.39	.37	1.08
Talisman	.31	.37	.95
Brewer's Gold	.41	<.01	n.d.
Backa	.83	.13	n.d.
Fuggle	.47	<.01	n.d.
Hallertauer	.43	.14	n.d.
Saazer	.32	.03	n.d.
Northern Brewer	.30	.04	n.d.
Cascade	.51	.14	2.11
Galena	.14	.04	.92
Cluster	.42	.47	.41

a. Ref. No. 2
b. Reported as % of oil.
c. Concentrated pellets.
n.d. = none detected.

aroma/taste must come from the water soluble oxygenated fraction of hop oil (13) as these compounds are often found in beer (4, 5, 7).

Humuladienone has been suggested to be important to kettle hop aroma in beer (5). Concentrations of from 34-72ppb humuladienone in beer and a sensory threshold of 100ppb for the compound in beer have been reported (5). Small amounts of humuladienone are formed by boiling humulene (a sesquiterpene typically comprising 25-45% of the aroma hop oils) for 90 minutes, and that the humuladienone concentration of hops increases with exposure to light and/or air (5). It has also been reported that the compound gives beer a "hop like" flavor at 100ppb (5).

Sandra and Verzele (14) have since reported looking for humuladienone in beer (claiming to have a detection limit of 1ppb) but failed to detect any. Tressl (7) has reported finding 10ppb humuladienone (along with an impressive list of other hop derived compounds) in a German beer. Peacock and Deinzer (4) could not detect humuladienone in a number of beers. Pickett (15) reports that boiling humulene in water results in almost exclusive oxidation to humulene epoxide II, with a few side products, but no detectable humuladienone. Also, irradiation of humulene with either a tungsten or mercury lamp (sensitized or unsensitized in solution) results in oxidation to humulene epoxide II. The mass spectrum of humuladienone and humulene epoxide II are almost indistinguishable (7) and their GC retention times on Carbowax 20M are quite close. It is likely humuladienone in beer was confused with humulene epoxide II (5).

Table I contains a number of hop derived compounds Peacock and Deinzer (4) have found in beers along with sensory threshold data which is available for them. They speculate that humulenol II (which is found in the mildly hoppy Cascade and European mixture beers, and to a greater extent in the hoppier Hallertauer beer but not in the less hoppy Cluster beer) is in part responsible for the hoppy aroma/taste of beer brewed with "aroma hops." Tressl (7) reports 1150ppb humulenol II in beer. Its sensory threshold in beer is reported as 500ppb (4). Peacock and Deinzer (4) speculate that humulenol II is only one humulene oxidation product that contributes to this flavor and that others such as humulol, humulene epoxides I and II, humulene diepoxide and possibly humuladienone are also flavor contributors, though less important. Tressl reports (7) 220ppb humulol, 125ppb humulene epoxide I, 40ppb humulene epoxide II and 10ppb humuladienone in beer. β-Eudesmol may also be important from consideration of the data in Table I. These compounds are in higher concentrations in the "hoppier" beers (Table I) and hops (Table III). Caryophyllene oxidation products are probably of little importance, as hops with high caryophyllene content are not considered to be aroma type hops. High humulene content is associated with "aroma hops."

Table III. Partial Analysis of Hop Oils[a]

Compound	Hallertauer	Hersbrucker	Tettnanger	Perle	Cascade	Northern Brewer	Cluster
Myrcene	12.28	8.32	23.92	14.92	46.93	28.16	39.71
Caryophyllene	15.41	12.56	14.27	15.70	6.56	11.20	9.77
Caryophyllene oxide	.18	.23	.05	.10	tr.	.01	.02
Caryolan-1-ol	.47	.05	.03	.31	.03	.31	.06
Humulene	46.65	29.18	41.56	48.87	17.41	36.88	20.08
Humulene Epoxide I	.40	.35	.05	.23	.08	.09	.12
Humulene Epoxide II	1.34	1.00	.38	.82	.55	.12	.18
Humuladienone	.08	.07	.03	.08	tr.	.04	--
Humulol	.14	.05	.11	.09	.03	.61	.08
Humulenol II	.50	.35	.14	.21	.23	.17	.16
Farnesene	--	--	6.43	--	6.95	--	--
Hop ether	.02	.02	< .01	.01	.03	.01	.01
Karahana ether	.02	.02	< .01	.01	.04	.01	.02
α-Eudesmol	.07	.94	--	--	--	--	--
β-Eudesmol	.21	.92	.03	.11	--	--	.02
T-Cadinol	.25	.41	.17	.27	.03	.34	.05
δ-Cadinol	.08	.22	.03	.01	.03	.08	--
α-Cadinol	.05	.28	.03	.02	--	--	.01
Humulene/ Caryophyllene ratio	3.0	2.3	2.9	3.1	2.7	3.3	2.1

a. Reported as % of oil. Oils were isolated and analyzed by procedure and methods of Ref. No. 2.

It has been noted recently (16) that hop aroma quality correlates well with the humulene/caryophyllene (H/C) ratio of the hop oil. This may just be an indirect way of saying high humulene concentration is important, but the results are of interest. Some hop varieties with an analysis of their more important constituents along with their (H/C) ratios are presented in Table III. Hallertauer, one of the most valuable aroma hops, has a high (H/C) ratio while Cluster (not considered an aroma hop) has a much lower (H/C) ratio. The other varieties on the list are considered to have aroma properties somewhere between Hallertauer and Cluster. Some other hops with low aroma properties are Galena (H/C = 2.2), Brewer's Gold (1.6) and Talisman (1.3).

High humulene levels in "aroma hops" have a number of implications: First, that higher concentrations of humulene oxidation products are likely in high humulene hops and that these oxidation products are going to increase in concentration during hop storage. Many brewers store hops for a year or so to improve their aroma properties. Tressl (17) has shown that concentrations of humulene oxidation products in hops go up dramatically with aging. Secondly, long wort boiling would increase oxidation of humulene to more water soluble compounds (15) and any residual humulene in the wort would likely be converted to humulene epoxide II during fermentation (6). Thus, even if humulene oxidation products are in low concentration in humulene rich hops, these compounds are likely to be found in beers brewed with such hops in higher concentrations than in beer brewed with hops with lesser amounts of humulene.

Another school of thought on "kettle hop" aroma in beer has been proposed by Tressl (6, 7). He reports finding the carotenoids β-ionone (3ppb) and β-damascenone (30ppb) in beer, and that the sensory threshold of β-ionone in water is 0.008ppb and that for β-damascenone in water is 0.009ppb (7). Peacock and Deinzer (18) have found the threshold of β-ionone in beer to be on the order of a few ppb. Tressl (6) also indicates that hop ether and karahana ether (reported in beer at 35 and 60ppb respectively), both with sensory thresholds of 5ppb in water, may be important to hop flavor. Both compounds have been identified in hops (17, 19).

These compounds may well have an effect on the hop aroma of beer but if they are responsible for the traditional "kettle hop" aroma, then the concentration of humulene in hops should not be so important. Further, from Table III the concentration of hop ether and karahana ether in hop oils does not correlate well with aroma quality. Tressl (17) has noted that the concentrations of these two ethers go up dramatically as hops age.

At present there is little agreement as to what compounds are responsible for "kettle hop" aroma. The theory of Peacock and Deinzer (4) that it is humulene oxidation products seems most in agreement with the traditional prejudices of brewers. Long hop storage, long wort boiling times and a high concentration of humulene in the hops would all help to maximize humulene oxidation products in the beer. However, Tressl's (6, 7) data on β-ionone, β-damascenone, hop ether and karahana ether cannot be ignored.

Literature Cited

1. Stevens, R. Chem. Rev. 1967, 67, 19.
2. Peacock, V.; Deinzer, M.; Likens, S.; Nickerson, G.; McGill, L. In press.
3. Micketts, R.; Lindsay, R. J. Food Protect. 1978, 41, 722.
4. Peacock, V.; Deinzer, M.; McGill, L.; Wrolstad, R. J. Agr. Food Chem. 1980, 28, 774.
5. Shimazi, T.; Hashimoto, N.; Kuroiwa, Y. Proc. Am. Soc. Brew. Chem. 1974, 9, 7.
6. Tressl, R.; Fendesack, F.; Köppler, H. ibid. 1979.
7. Tressl, R.; Friese, L.; Fendesack, F.; Köppler, H. J. Agr. Food Chem. 1978, 26, 1422.
8. Meilgaard, M. MBAA Tech. Quart. 1975, 12, 151.
9. Meilgaard, M. J. Am. Soc. Brew. Chem. 1980, 38, 99.
10. Likens, S.; Nickerson, G. J. Agr. Food Chem. 1967, 15, 525.
11. Howard, G. Brewer's Guardian 1965, 3, 17.
12. Maule, D. J. Inst. Brew. 1967, 73, 351.
13. Buttery, R.; Ling, L. J. Agr. Food Chem. 1967, 15, 531.
14. Sandra, P.; Verzele, M. Proc. Eur. Brew. Con. 1975, 109.
15. Pickett, J.; Sharpe, F.; Peppard, T. Chem. Ind. London 1977, 30.
16. Kralj, D. Eur. Brew. Con. Hops Workshop, 1978.
17. Tressl, R.; Friese, L.; Fendesack, F.; Köppler, H. J. Agr. Food Chem. 1978, 26, 1426.
18. Unpublished work.
19. Naya, Y.; Kotake, M. Bull. Chem. Soc. Jap. 1972, 45, 2287.

RECEIVED April 3, 1981.

11

Volatile Constituents of Pumpkins

THOMAS H. PARLIMENT, MICHAEL G. KOLOR, DONALD J. RIZZO, and DENISE HERZING-GIORDANO

General Foods Corporation, Technical Center, 250 North Street, White Plains, NY

Pumpkins (Cucurbita pepo) are a member of the family Cucurbitaceae which includes squashes, melons, pumpkins and cucumbers. Pumpkins are of New World origin and are believed to have been important in the diet of aboriginal Americans. The flavor composition of the Cucurbits has received little attention though other members of the Cucurbitaceae have been investigated. For example, the carbonyl compounds of fresh cucumbers (Cucumis sativus) were studied almost 20 years ago (1). More recently, the total volatile composition of distilled cucumbers was investigated by a number of groups (2,3). The steam volatile compounds of muskmelons (Cucumis melo) have also received attention (4,5,6) as have those of watermelons (Citrullus vulgaris) (7). The most striking feature in the volatile components of all these fruits is the variety and magnitude of the nine-carbon compounds. These compounds include nonanol, nonanal, various nonenols, nonadienols, nonenals and nonadienals.

The purpose of this study is to identify the major volatile components of freshly cooked pumpkins and to compare these to the volatiles of commercially canned pumpkins.

0097-6156/81/0170-0129$05.00/0

EXPERIMENTAL

Sample Preparation

In a typical experiment, 4 kg of fresh, locally grown pumpkins (variety: Connecticut field) were homogenized with 2 L of water and this slurry was placed in a 12 L three-necked flask. The sample was heated at reflux for 3 hours using a modified Likens-Nickerson distillation head (8). The distillation was performed at atmospheric pressure and diethyl ether was used as the extracting solvent. The ethereal concentrate was dried over anhydrous sodium sulfate, filtered and concentrated using a low hold-up distillation column by the usual procedure.

For our studies of canned pumpkin, 1.2 kg of commercially canned pumpkin (ingredient line reads: pumpkin, water) were combined with 800 ml water and the slurry was worked up in the same fashion as was the fresh pumpkin.

Separation and Identification

Separation of the mixture was accomplished by gas-liquid chromatography (GLC) in a Perkin-Elmer Model Sigma II (Perkin-Elmer Corporation, South Norwalk, Connecticut) using a capillary column. The chromatograph had an all glass linear splitter (split 80:1) and a helium linear velocity of 23 cm/sec. A Perkin-Elmer fused silica column was used which measured 0.235 mm id, x 25 m long and which was coated with OV-101 (methyl silicone) liquid phase. The column was held 4 min at 45°C then programmed at 3°C/min to 120°C. Mass spectra were obtained using tandem gas chromatography/mass spectrometry. The column effluent was passed into the ion source of a DuPont Model 21-491 mass spectrometer (DuPont Instrument Division, Wilmington, Delaware). Mass spectra were obtained at 70 eV and a source temperature of 200°C.

Component identification was accomplished by comparison of mass spectra and gas chromatographic retention times of the isolated materials to that of known standards which were commercially available or in our reference library. Retention indices relative to ethyl esters (9) were calculated for all peaks and compared to those of authentic compounds or to the values listed in our files. Sample quantitation was performed on a Perkin-Elmer Sigma 10 data system. Percent of each component (on a solvent free basis) was calculated, using response factors of 1.

RESULTS AND DISCUSSION

Identification of the major components of cooked pumpkins is presented in Table I and the gas chromatogram is shown in Figure 1.

TABLE I VOLATILE COMPONENTS OF FRESHLY COOKED PUMPKINS

PEAK NUMBER(1)	IDENTITY	% COMPOSITION
1	Diacetyl	6.2
2	Ethyl Acetate	0.9
3	Ethyl Furan	2.6
4	3-Pentanone	1.3
5	Pyridine	2.7
6	2-Pentenal	1.5
7	1-Pentanol	1.8
8	n-Hexanal	11.1
9	Furfural	1.5
10	2-Methyl-2-Pentenal	0.4
11	2-Hexenal	17.8
12	cis-3-Hexenol	20.6
13	n-Hexanol	21.6
14	Methional	0.5
15	2-Heptanone	0.2
16	2-Heptenal	0.4
17	Benzaldehyde	0.4
18	1-Octene-3-ol	0.5
19	Pentyl Furan	0.8
20	Phenyl Acetaldehyde	0.6
21	Methyl Thiophene Carboxyaldehyde	0.6
22	Phenyl Ethanol	0.7

(1) As Indicated in Figure 1

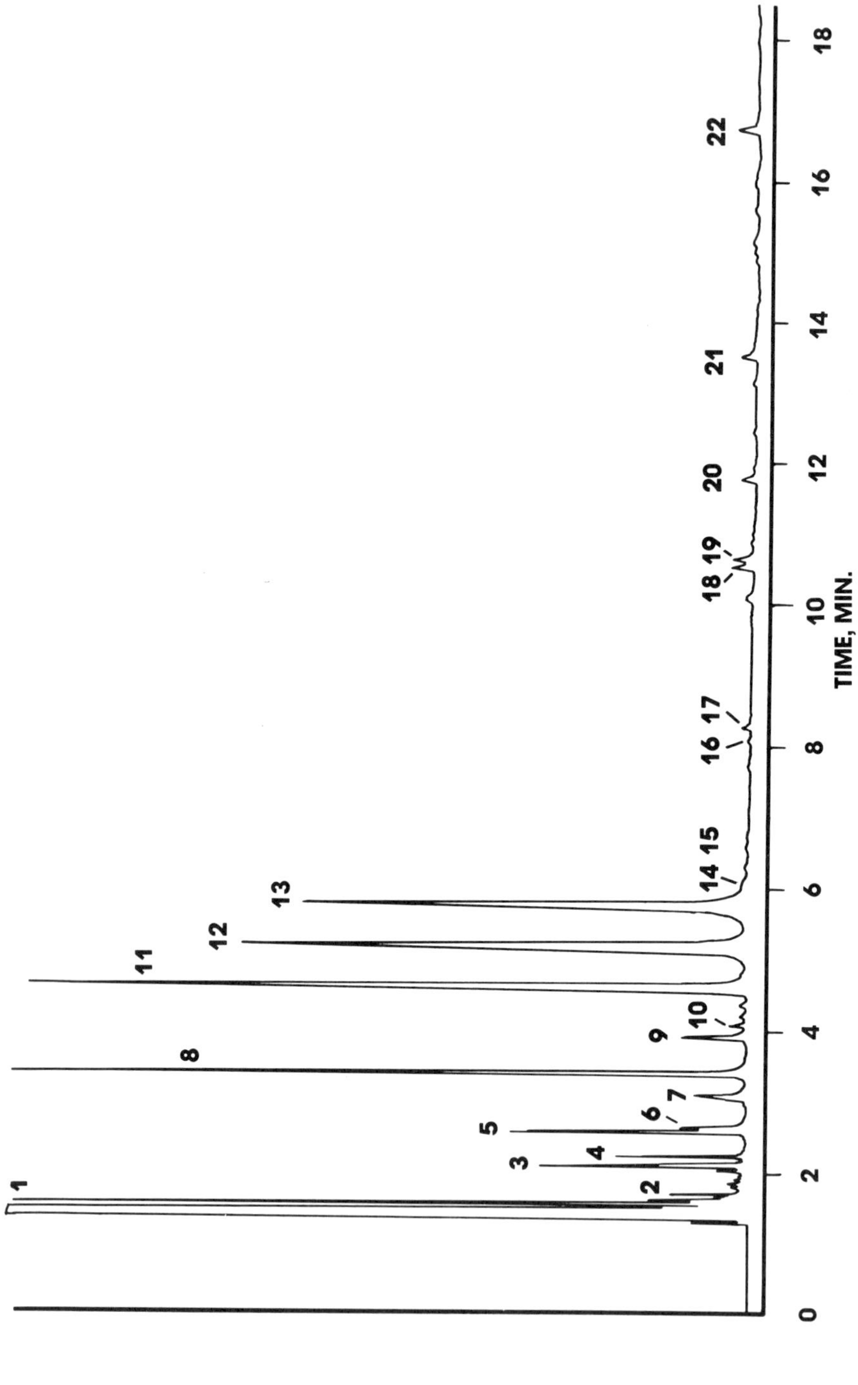

Figure 1. Gas chromatogram of volatile components of freshly cooked pumpkins

The surprising feature of this fruit is the quantity of six-carbon alcohols and aldehydes. Taken as a group, these compounds constitute over 70% of the total volatiles. We had fully expected to find a variety of nine-carbon compounds in our essence and we specifically looked for them since they are the major volatiles in other Cucurbitaceae. For example, nonanal, 2-nonenal, 6-nonenal and 2,6-nonadienal are some of the major components of cucumber (3) and 6-nonenol is the major compound in frozen muskmelon (5). None of these compounds were found; if present, they are below the 1% level.

The identities of the major volatile components of commercial canned pumpkin are presented in Table II and the gas chromatogram is shown in Figure 2.

TABLE II VOLATILE COMPONENTS OF COMMERCIALLY CANNED PUMPKINS

PEAK NUMBER(2)	IDENTITY	% COMPOSITION
1	Diacetyl	8.7
2	Ethyl Acetate	2.0
3	iso Valeraldehyde	4.1
4	2-Methyl Butanal	16.7
5	2,3-Pentanedione	2.4
6	Pyridine	15.1
7	Toluene	1.1
8	2-Methyl Tetrahydrofuran-3-one	1.6
9	Furfural	15.0
10	2-Methyl-2-Pentenal	3.2
11	cis-3-Hexenol	3.5
12	Methyl Furyl Ketone	0.5
13	2,5-Dimethyl Pyrazine	0.9
14	Dimethyl Trisulfide	0.9
15	Phenyl Acetaldehyde	2.2

(2) As Indicated in Figure 2.

The compounds here are dramatically different from those in fresh cooked pumpkin. This is best shown in Table III where the 10 major compounds of each sample are compared.

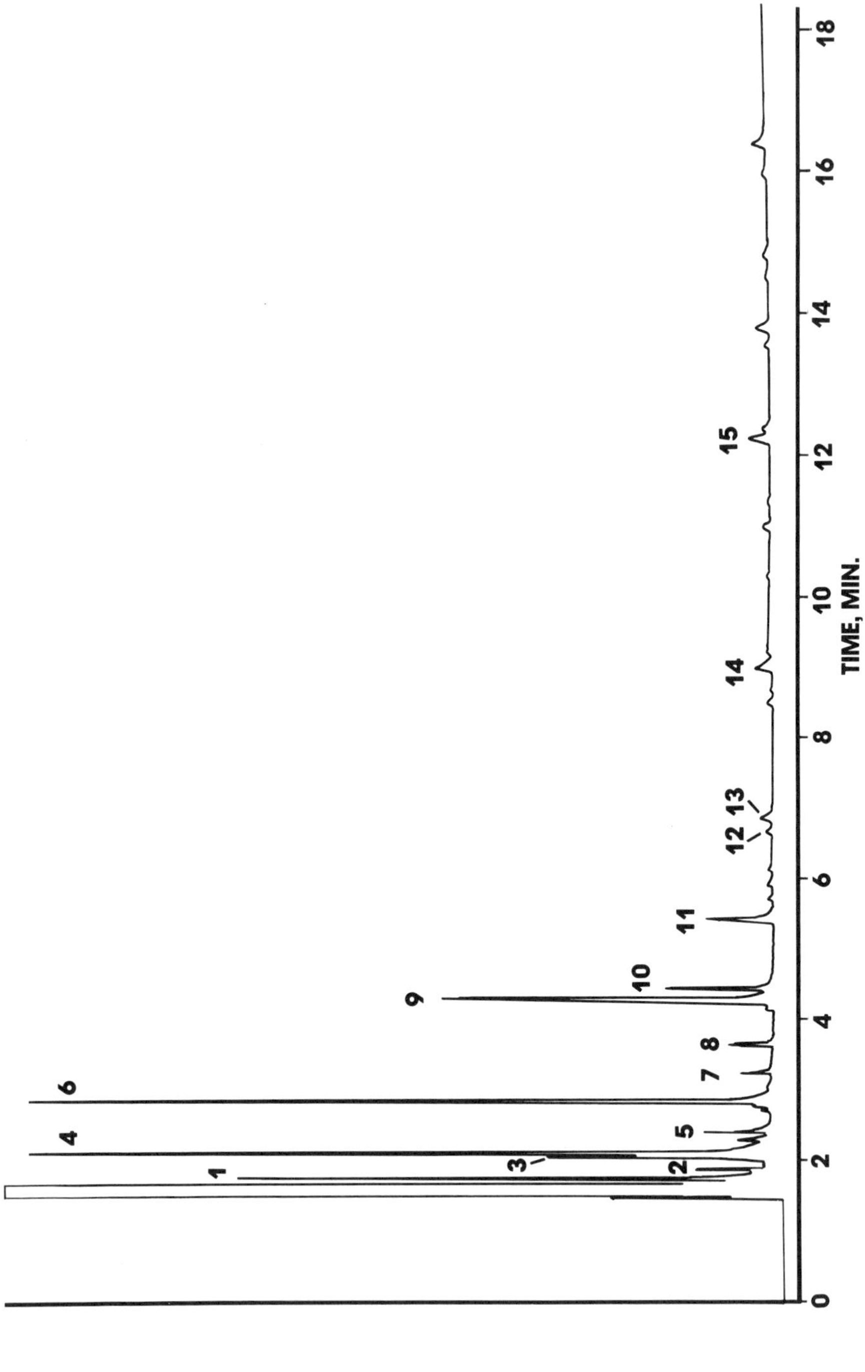

Figure 2. Gas chromatogram of volatile components of canned pumpkins

TABLE III COMPARISON OF COMPONENTS IN PUMPKINS

	FRESH COOKED	
RANK	IDENTITY	% COMPOSITION
1	n-Hexanol	21.2
2	cis-3-Hexenol	20.6
3	2-Hexenal	17.8
4	n-Hexanal	11.1
5	Diacetyl	6.2
6	Pyridine	2.7
7	Ethyl Furan	2.6
8	Furfural	1.5
9	3-Pentanone	1.3
10	Ethyl Acetate	0.9

	CANNED	
RANK	IDENTITY	% COMPOSITION
1	2-Methyl Butanal	16.7
2	Pyridine	15.1
3	Furfural	15.0
4	Diacetyl	8.7
5	iso Valeraldehyde	4.1
6	2-Methyl-2-Pentenal	3.2
7	cis-3-Hexenol	3.5
8	2,3-Pentanedione	2.4
9	Phenyl Acetaldehyde	2.2
10	Ethyl Acetate	2.0

Virtually all of the six-carbon aldehydes and alcohols found in freshly cooked pumpkin have been lost. Furfural has increased significantly and 2-methyl tetrahydrofuran-3-one has appeared. These changes are likely due to the severe thermal processing canned pumpkins must receive. Pumpkins are a low acid food which heat primarily by conduction and as such require extended processing times at 250°F or higher. Furfural is a well known constituent of thermally decomposed carbohydrates, and the furanone as well as the furyl-ketone have been reported in numerous cooked food products (10). 2-Methyl butanal and iso valeraldehyde are also common constituents of processed foods which can arise from the Strecker Degradation of isoleucine and leucine, respectively. Similarly, model system studies have shown that dimethyl pyrazines result from heating sugar and amino acids together. Thermal processing doubtless results in the other qualitative and quantitative differences which are evident when freshly cooked and canned pumpkins are compared. Another interesting feature of both freshly cooked and canned pumpkins is the virtual absence of esters. This is unusual since the volatile flavor components of most fruits are rich in esters, yet all we found were quite low levels of ethyl acetate.

LITERATURE CITED

1. Forss, D.A., Dunstone, E.A., Ramshaw, E.H., Stark, W., *J. Food Sci*. 1962, *27* 90.
2. Hatanaka, A., Kajiwara, T., Harada, T., *Phytochemistry* 1975, *14* 2589.
3. Kemp, T.R., Knavel, D.E., Stoltz, L.P., *J. Agr. Food Chem*. 1974, *22* 717.
4. Kemp, T.R., Stoltz, L.P., Knavel, D.E., *J. Agr. Food Chem*. 1972, *20* 196.
5. Kemp, T.R., Knavel, D.E., Stoltz, L.P., *Phytochemistry* 1973, *12* 2921.
6. Yabumoto, K., Jennings, W.G., *J. Food Sci*. 1977, *42* 32.
7. Kemp, T.R., Knavel, D.E., Stoltz, L.P., Lundin, R.E., *Phytochemistry* 1974 , *13* 1167.
8. Schultz, T.H., Flath, R.A., Mon, T.R., Eggling, S.B., Teranishi, R., *J. Agr. Food Chem*. 1977, *25* 446.
9. van den Dool, H., Kratz, P., *J. Chrom*. 1963, *11* 463.
10. van Straten, S., Volatile Compounds in Food. 4th. Edition, including Supplement 1 and 2. Central Institute for Nutrition and Food Research, Zeist, Netherlands, 1978.

RECEIVED April 3, 1981.

12

Volatiles from Red Pepper (*Capsicum* spp.)

URS KELLER
Firmenich SA, Geneva, Switzerland

ROBERT A. FLATH, THOMAS R. MON, and ROY TERANISHI
Western Regional Research Center, SEA-AR, U.S. Department of Agriculture, Berkeley, CA 94710

Fruit of the genus Capsicum has been included in the diet of Western Hemisphere peoples since long before the European discovery of the Americas (1,2). After European contact with the New World was made, Capsicum plants were dispersed through much of the remaining world, especially to sub-tropical and tropical regions. The fruits, both fresh and dried, are used as spices and as staples in the diet. An obvious characteristic of the fruits is their "hotness" or "bite" when eaten. This character is quite variable among the members of the genus, with some varieties exhibiting such hotness that they can only be added in small quantities as a spice to food dishes, whereas others exhibit no discernable bite at all. In addition, the fruits have rather distinctive flavor characters, making them very popular components of certain ethnic diets, especially when the hotness is not overpowering.

Botanical classification of Capsicum members is rather confused (2,3,4), but recent sources include at least two major species, C. annuum and C. frutescens. As a rough generalization, the larger peppers tend to be C. annuum. These frequently are somewhat lower in hotness, and this characteristic may be entirely lacking in certain varieties. C. frutescens tends to include the smaller, hotter peppers. Certain botanical characteristics are also considered when classifying Capsicum plants into one or another species. Superimposed on the botanical classifications are the assortment of trivial names for individual varieties. The Hot Chilis, Jalapenos, and Tabasco peppers are usually considered to be C. frutescens, while Sweet Chilis, Bell Peppers, Paprika, and Cayenne are included among the C. annuum varieties. The term "red pepper" refers to the typical color that most present-day pepper varieties attain when permitted to ripen thoroughly, although yellow colorations are not unknown, and some early references to violet and white fruits may be found (1).

The hotness of peppers (unrelated to black pepper [Piper nigrum L.]) is due to various capsaicinoids in the fruit.

0097-6156/81/0170-0137$05.00/0

Capsaicin, the major capsaicinoid, is the N-vanillyl amide of 8-methyl-6-nonenoic acid. Related dihydro-, homo-, and dihydrohomo- isomers have also been reported. Most research efforts on red pepper have been concerned with determining the capsaicinoid makeup and concentration in various red pepper varieties and products (5-8).

Several groups of workers have examined the volatiles makeup of certain pepper varieties. The first of these were Buttery *et al.* (9), who isolated and identified 2-methoxy-3-isobutylpyrazine from fresh bell peppers. They determined its odor threshold in water to be 2 parts in 10^{12} parts of water. Haymond and Aurand (10) identified twenty-four components of fresh Tabasco peppers, including a number of esters, and decided that all contribute to Tabasco aroma. Murray and Whitfield (11), in their extensive survey of 3-alkyl-2-methoxypyrazines in raw vegetables, reported finding 3-isobutyl-, 3-isopropyl-, and 3-(*sec*-butyl)-2-methoxypyrazines in red, or chili peppers. Combining gc effluent sniffing and gc-ms selected ion monitoring, Huffman *et al.* (12) detected 3-isobutyl-2-methoxypyrazine in both fresh and processed Jalapeno pepper volatiles. They attributed most of the fresh Jalapeno aroma to the presence of the bell pepper pyrazine. In a recent communication on *Capsicum* volatiles concentration, Teranishi *et al.* (13) reported isolating from the oleoresin the methyl esters of lauric, palmitic, and stearic acid, as well as free acetic acid, along with traces of several C_4-C_6 acids.

Experimental

Materials. Oleoresin of *Capsicum*, African type, was purchased in several lots from Kalsec, Inc., Kalamazoo, MI. This material is a viscous deep orange-red liquid/semisolid residue obtained by hexane extraction of ground, dried ripe *C. frutescens* peppers. Commercially, it is primarily used as a natural source of "hotness" and flavor for foods and for animal repellent formulations. This material was used in the initial analytical studies.

Sweet chili miscella (solution of extracted material) was also obtained from Kalsec, Inc. This material is also hexane-extracted, but from sweet chilis (*C. annuum* [14]). It is much less "hot" than is the oleoresin *Capsicum*, and is mostly used for flavoring and coloring foods. Normally the extracting solvent is stripped from the extract by the supplier before shipment. In order to minimize volatiles loss during this step, the hexane content of the extract was only reduced to *ca* 20% of the total volume before shipment.

Fresh Jalapeno peppers (*C. frutescens*) were obtained from a local market. The cultivar is unknown.

Distilled water was used in all volatiles concentration procedures. All solvents were distilled in glass equipment immediately before use.

An antioxidant (1,3,5-trimethyl-2,4,6-tris[3,5-di-*tert*-butyl-4-hydroxybenzyl]benzene; Antioxidant 330, Ethyl Corporation) was added to all fractions and concentrates which were to be stored for any length of time.

Concentrates Preparation. Volatiles concentrates were prepared from starting materials with a modified Likens and Nickerson steam distillation-extraction head (15) operated either at 1 atm. (ether solvent) or at reduced pressure (50-55mm; heptane solvent).

Oleoresin *Capsicum* - In a typical atmospheric pressure run, a quantity (1.48 kg) of oleoresin was combined in a 12 l round-bottomed flask with distilled water (6 l). Ether (200 ml) was used as extraction solvent. The steam distillation-extraction process was run for 4 hr. The resulting ether solution was dried (anhyd. sodium sulfate) and the ether distilled, leaving 2.49 g of pepper components (containing 10% ether); 0.15% yield after correcting for residual solvent.

In a typical reduced pressure run (55 mm), oleoresin (1.92 kg) and distilled water (6 l) yielded 3.38 g (0.176% yield) of concentrate when heptane was used as the extraction solvent. Two 4-hr extraction periods were used in each run for concentrate preparation at reduced pressure.

Sweet Chili Miscella - Concentrates were prepared at reduced pressure only. Initial attempts to use the miscella as obtained from the supplier were unsuccessful. When the miscella was combined with distilled water and the mixture was heated with stirring, an intractable gel formed. Subsequently, a hexane fraction was first distilled from the starting material at reduced pressure. Pepper volatiles entrained in this distillate were later reclaimed by careful fractional redistillation of the hexane fraction. The residue from this fractional distillation was combined with the heptane-extracted material obtained with the steam distillation-extraction head. In a typical run, 5.28 kg of miscella yielded 1.8 l of hexane distillate. The miscella residue was combined with 12 l of distilled water, and three 4-hr extraction runs were made at 50 mm pressure. On redistillation of the 1.8 l hexane distillate, 0.55 g of residue was obtained. The combined yield from the entire process was 1.05 g (0.0257%).

Fresh Jalapeno Peppers - In a typical run, the peppers were quartered (5 kg) and added to boiling distilled water (5 l). The distillation-extraction process was run for four 2-hr periods, using freshly-distilled ether. A total yield of 0.143 g (0.0029%) was obtained after removal of solvent.

Concentrates Fractionation. Oleoresin Capsicum - Conventional vacuum fractional distillation, aluminum oxide adsorption chromatography, and preparative gas chromatography were applied in preliminary efforts to fractionate oleoresin Capsicum volatiles.

Sweet Chili Miscella Volatiles - A portion (5.00 g) of sweet chili volatiles in a 40-ml pentane solution was separated into neutral, acid, base fractions by successive extraction with saturated sodium bicarbonate solution and 6N hydrochloric acid solution. The neutral organic layer, after drying and solvent removal, yielded 4.00 g of organic residue (80% yield). The bicarbonate extract, after acidification and solvent extraction, afforded 0.110 g (2.2%) of free acids. The hydrochloric acid solution, on basification with 5N sodium hydroxide solution and solvent extraction, yielded 0.080 g (1.6%) of organic material.

Component Separation and Identification. Hewlett Packard 5830A and 5840A gas chromatographs fitted with glass, fused silica, or stainless steel capillary columns of various inside diameters were used for separation of the volatiles fractions. Effluent splitters were used on occasion to permit gc effluent sniffing. Methyl silicone oil was the stationary phase of choice for most of the work, although Tween 20 and Carbowax 20M were used occasionally.

A quadrupole-type gc-ms (16) was used for component identification in nearly all instances. A few components were isolated by preparative gc for infrared and nuclear magnetic resonance spectrometric examination. The retention behaviors of all tentatively-identified constituents were checked, using authentic samples, in order to verify the ms interpretations.

Sensory Evaluation. Odor thresholds were determined as previously described, using Teflon squeeze bottles for panel testing (17).

Results and Discussion

A summary of the compounds identified in various pepper samples is provided in Table I. Many of the identifications were obtained with volatiles fractions from oleoresin Capsicum.

The three major compounds in the capsaicinoid fraction of most "hot" Capsicum fruits are capsiacin, dihydrocapsaicin, and nordihydrocapsaicin (6,7,8). These compounds are the N-vanillyl amides of 8-methyl-6-nonenoic acid, 8-methylnonanoic acid, and 7-methyloctanoic acid, respectively. Inspection of Table I reveals that these three acids were found in the oleo-

Table I. Compounds Identified in Pepper Volatiles by Gc-Ms.

Alcohols
cyclopentanol a
2-methylbutan-1-ol b
2-methylbutan-2-ol b
3-methylbutan-1-ol a
1-pentanol a
2,3-butanediol b
trans-2-hexen-1-ol a
cis-3-hexen-1-ol a, c, e
2-methylpentan-2-ol b
3-methylpentan-3-ol b
4-methylpentan-1-ol a, c
1-hexanol a, b, e
2-hexanol a, b
3-hexanol b
linalool a, b, c
terpinen-4-ol a, b
alpha-terpineol b, c
Aldehydes and Ketones
2-butanone b
2-methylbutanal a, b
3-methylbutanal a, b
cyclohexanone b
4-methyl-3-penten-2-one a, b
n-hexanal a, b, e
2-hexanone b
benzaldehyde a, e
2-acetylfuran b
5-methyl-2-furfural a, b
2-heptanone a
2-octanone a
para-methylacetophenone a
carvone a
camphor a
thujone a
iso-thujone a
2-undecanone a
beta-ionone c
geranylacetone a
Carboxylic Acids
acetic a
2-methylpropionic b
2-methylbutyric a, b
3-methylbutyric b
pentanoic a
4-methylpentanoic a, b
hexanoic a, b
heptanoic a

Table I. Compounds Identified in Pepper Volatiles by Gc-Ms. Continued

- 2-octenoic *a*
- octanoic *a*, *b*
- 7-methyloctanoic *a*
- nonanoic *a*
- 2-decenoic *a*
- 8-methylnonanoic *a*

Esters
- methyl pentanoate *a*
- methyl hexanoate *a*
- ethyl 3-methylbutyrate *b*
- methyl heptanoate *a*
- methyl phenylacetate *a*
- methyl octanoate *a*
- *beta*-phenylethyl acetate *a*
- methyl *beta*-phenylpropionate *a*
- ethyl octanoate *a*
- methyl nonanoate *a*
- methyl 8-methyl-6-nonenoate *a*
- 4-methylpentyl 2-methylbutyrate *a*, *c*, *e*
- 4-methylpentyl 3-methylbutyrate *a*, *c*
- methyl 8-methylnonanoate *a*
- methyl decanoate *a*
- methyl dodecanoate *a*
- ethyl dodecanoate *a*
- methyl tetradecanoate *a*, *h*
- ethyl tetradecanoate *a*
- methyl hexadecanoate *a*, *h*
- methyl octadecanoate *a*, *h*

Pyrazines
- 2,3-dimethylpyrazine *b*
- 2,3,5-trimethylpyrazine *b*
- 2-methyl-5-ethylpyrazine *b*
- 2,3-dimethyl-5-ethylpyrazine *b*
- tetramethylpyrazine *b*
- 2-methoxy-3-isobutylpyrazine *a*, *b*, *d*, *f*, *g*

Terpene Hydrocarbons
- *para*-cymene *b*
- camphene *b*
- *delta*-3-carene *a*, *b*
- limonene *a*, *b*
- myrcene *a*, *b*
- *alpha*-phellandrene *a*, *b*
- *alpha*-pinene *b*
- *beta*-pinene *a*, *b*
- sabinene *b*
- *gamma*-terpinene *b*
- terpinolene *b*
- *alpha*-thujene *b*
- caryophyllene *a*, *b*
- *alpha*-copaene *b*

Miscellaneous
- toluene *b*, *e*
- *para*-xylene *a*, *b*
- octane *b*
- 2-pentylfuran *a*, *b*
- 2-pentylpyridine *b*
- 1,8-cineole *a*
- eugenol *b*
- pentadecane *a*
- hexadecane *a*
- heptadecane *a*

a - Oleoresin, *Capsicum frutescens*; *b* - Sweet chili, *C. annuum*;
c - Fresh Jalapenos, *C. frutescens*; *d* - Buttery *et al.*, 1969;
e - Haymond and Aurand, 1971; *f* - Murray and Whitfield, 1975;
g - Huffman *et al.*, 1978; *h* - Teranishi *et al.*, 1980.

resin Capsicum volatiles, as free acids and/or as their methyl esters.

Different lots of oleoresin Capsicum obtained at different times varied considerably in aroma character and intensity. This was primarily due to variations in the sources of starting material for preparation of the oleoresin, and was one of the major reasons for selecting the sweet chili material for subsequent research efforts. The sweet chilis are grown in New Mexico.

Numerous attempts were made to fractionate the oleoresin Capsicum volatiles in such a manner that a peppery aroma would be concentrated in one fraction. Vacuum fractional distillation yielded four fractions plus a residue, none of which exhibited a peppery aroma. The first fraction eluted with pentane from a deactivated (15% water by weight) alumina adsorption column had a peppery aroma, but on rechromatographing the fraction on more active alumina, the pepper aroma was lost. The aroma was also mostly lost when preparative gas chromatography was applied in an attempt to separate the peppery alumina column fraction. Gc fractions which retained some pepper-like aroma lost this aroma after a few hours, even at low temperatures. Some success was achieved with larger bore glass capillary columns; components with odors reminiscent of the pepper aroma partly survived separation in such columns, permitting gc sniffing of the effluent. However, attempts to collect several components were unsuccessful.

The neutral, acidic, and basic fractions of the sweet chili volatiles concentrate were examined in some detail, and efforts were made to correlate the various fractions' odor thresholds and odor character with the components found in each.

The neutral material, representing 80% of the original sweet chili volatiles, consist largely of monoterpene hydrocarbons. Sabinene (15.0%), delta-3-carene (13.4%), alpha-pinene (9.7%), limonene (9.2%), beta-pinene (7.3%), and several other minor monoterpenes comprise 63% of the neutral fraction. A number of aliphatic hydrocarbon solvent components were detected, along with considerable residual heptane (13%). Only two sesquiterpenes, alpha-copaene and caryophyllene were identified. Both are present at low concentrations. A variety of lower molecular weight aliphatic aldehydes, ketones, and alcohols were also present in small amounts. Approximately 92% of the neutral material gc peak area has been identified. Assuming the odor unit concept to be valid above threshold levels (18), alpha-pinene and limonene together provide about one-third of the total neutral fraction's aroma. The odor threshold of the total fraction in water is 11 parts per billion.

The acidic fraction is 2.2% of the total volatiles by weight. While several of the acids were identified by direct gc-ms of the free acid fraction, most identifications as well as quantitative data were obtained after conversion to the methyl esters. Three lower aliphatic acids, 2-methylbutyric (42.8%), 3-methylbutyric (25.3%), and 2-methylpropionic (4.8%) make up 78% of the acid fraction. In addition, traces of 4-methylpentanoic, *n*-hexanoic, and *n*-octanoic acids were detected. Several non-acid constituents were also found in the methyl ester fraction, including trimethylpyrazine (1.1%), tetramethylpyrazine (1.7%), traces of several ketones and 3-methyl-3-pentanol. Eighty-two percent of the total chromatographic peak area has been identified. A major contributor to the aroma of the free acid material is 2-methylpropionic acid (odor threshold = 50 ppb [19]). On an odor unit basis, this acid may be shown to provide approximately one-half of the total odor units of the fraction. The measured threshold of this fraction (free acid) is 255 ppb. Considering that this is roughly 20 times the threshold of the neutral fraction, and taking into account the small amount of acidic material in the total volatiles, the acids are not thought to be significant contributors to the aroma of the total sweet chili volatiles.

The basic fraction comprises only 1.6% of the total sweet chili volatiles. Tetramethylpyrazine makes up 68% of the fraction; in addition, lesser amounts of several other pyrazines were identified, including 2,3-dimethyl-5-ethyl-, 2-methyl-5-ethyl-, 2,3,5-trimethyl-, 2,3-dimethyl- and 2-methoxy-3-isobutylpyrazine, listed in descending order of concentration. Greater than 75% of the basic material consists of pyrazines, some of which have not been fully characterized. A rather low threshold of 1.4 ppb was determined for this fraction. Considering that the bell pepper pyrazine's measured concentration in this basic fraction is 0.23% and its odor threshold is 2 parts per trillion (10^{12}), this pyrazine might be expected to be the major source of the basic fraction's aroma (tetramethylpyrazine's odor threshold is reported to be 10 ppm [20]). However, since a considerable margin of error exists in quantitating such small gas-chromatographic peaks, and since threshold determination is not without its sources of error, such a conclusion is not necessarily valid. Indeed, the aroma of the total sweet chili miscella volatiles fraction, and for that matter, of the sweet chili miscella itself, is quite different from that of the bell pepper pyrazine alone. A dried cooked vegetable tissue odor reminiscent of ground paprika predominates at full strength on a blotter strip, but when a dilute (*ca* 1:100) pentane solution of the sweet chili miscella is sampled in similar fashion, the cooked odor becomes subdued, and a distinct floral note is detected, along with an underlying bell pepper aroma.

Some 16% of the original sweet chili volatiles material is not accounted for by the sum of the neutral, acid, and basic fractions. This is thought to be primarily due to the presence of considerable 2,3-butanediol in the starting volatiles concentrate. This material is quite water soluble, and would be retained in the aqueous phases during separation into three fractions. The diol was detected by gc-ms in the original volatiles material, but was not found in any of the three fractions.

Numerous minor constituents remain unidentified, especially in the neutral and basic fractions. Sniffing runs have indicated that quantitatively-minor components may still provide considerable aroma character, so these minor constituents cannot necessarily be ignored.

The fresh Jalapeno pepper volatiles concentrate has a pleasant floral aroma, along with an appreciable bell pepper character. The concentrate was not examined in detail, but gc-ms data indicates the presence of numerous higher aliphatic esters, including the two which were synthesized and identified - the 4-methylpentyl esters of 2-methylbutyric (5.1%) and 3-methylbutyric (2.2%) acid. Attempts to identify 2-methoxy-3-isobutylpyrazine were frustrated by considerable component overlap in the gc-ms run.

Summary

Volatiles from red pepper or chilis (*Capsicum annuum* and *C. frutescens*) were isolated by steam distillation-solvent extraction methods. Compounds identified by gas chromatography-mass spectrometry are: 17 alcohols, 20 carbonyl compounds, 14 carboxylic acids, 21 esters, 6 pyrazines, 14 terpene hydrocarbons, and 10 miscellaneous compounds. Only ten of these components have been reported previously in volatiles from peppers. Although the odor threshold method was applied, the characteristic red pepper odor was not fully correlated with identified constituents. The green bell pepper compound, 2-methoxy-3-isobutylpyrazine, was found in both *C. annuum* and *C. frutescens*.

Literature Cited

1. Hedrick, U. P., Ed. "Sturtevant's Notes on Edible Plants"; Twenty-Seventh Annual Report, Vol. 2, Part II, State of New York - Department of Agriculture; J. B. Lyon Company, State Printers: Albany, NY, 1919; p 134-140.
2. Bailey, L. H.; Bailey, E. Z. "Hortus Third"; Macmillan Publishing Co., Inc.: New York, 1976; p 219.
3. Bailey, L. H. "Manual of Cultivated Plants"; Macmillan Publishing Co., Inc.: New York, 1949; p 873.

4. Arctander, S. "Perfumery and Flavoring Materials of Natural Origin"; S. Arctander, Publ.: Elizabeth, NJ, 1960; p 122-124.
5. Todd, P. H., Jr.; Bensinger, M. G.; Biftu, T. J. Food Sci. 1977, 42, 660-665, 680.
6. Sticher, O.; Soldati, F.; Joshi, R. K. J. Chromatogr. 1978, 166, 221-231.
7. Iwai, K.; Suzuki, T.; Fujiwake, H. Agric. Biol. Chem. (Japan) 1979, 43, 2493-2498.
8. Jurenitsch, J.; David, M.; Heresch, F.; Kubelka, W. Planta Medica 1979, 36, 61-67.
9. Buttery, R. G.; Seifert, R. M.; Lundin, R. E.; Guadagni, D. G.; Ling, L. C. Chem & Ind. (London) 1969, 490-491.
10. Hayman, L. W.; Aurand, L. W. J. Agric. Food Chem. 1971, 19, 1131-1134.
11. Murray, K. E.; Whitfield, F. B. J. Sci. Fd. Agric. 1975, 26, 973-986.
12. Huffman, V. L.; Schadle, E. R.; Villalon, B.; Burns, E. E. J. Food Sci. 1978, 43, 1809-1811.
13. Teranishi, R.; Keller, U.; Flath, R. A.; Mon, T. R. J. Agric. Food Chem. 1980, 28, 156-157.
14. Cripps, H. D. Kalsec, Inc.; private communication 1979.
15. Schultz, T. H.; Flath, R. A.; Mon, T. R.; Eggling, S. B.; Teranishi, R. J. Agric. Food Chem. 1977, 25, 446-449.
16. Flath, R. A.; Forrey, R. R. J. Agric. Food Chem. 1977, 25, 103-109.
17. Buttery, R. G.; Seifert, R. M.; Guadagni, D. G.; Black, D. R.; Ling, L. C. J. Agric. Food Chem. 1968, 16, 1009-1015.
18. Guadagni, D. G.; Buttery, R. G.; Harris, J. J. Sci. Fd. Agric. 1966, 17, 142-144.
19. Holluta, J. Gas Wasserfach 1960, 101, 1018-1023.
20. Koehler, P. E.; Mason, M. E.; Odell, G. V. J. Food Sci. 1971, 36, 816-818.

RECEIVED April 3, 1981

13

Peanut Quality: Its Relationship to Volatile Compounds—A Review

HAROLD E. PATTEE and JOHN A. SINGLETON

USDA, SEA, AR, Mid-Atlantic Area and Department of Botany, North Carolina State University, Raleigh, NC 27650

The quality of peanuts depends on many factors, such as cultivar, maturity, cultural practices, curing temperature, and storage conditions. Most of these factors have also been shown to affect the volatile profile of peanuts, some qualitatively and other quantitatively. The volatiles of raw peanuts are produced by the action of the enzyme lipoxygenase, and the amounts produced reflect the various factors influencing lipoxygenase activity. The information presented can have a significant impact on future modifications of the grades and standards for marketing peanuts.

Quality and quantity of any crop determine its price in the market place. Although quantity is readily measurable, quantity in units of weight or volume, quality is a less readily measurable attribute. Quality pertains to the characteristics of the commodity and varies somewhat according to the perceptions of different individuals. Nevertheless, grades and standards can and have been established for numerous products to describe their quality characteristics. Many current grades and standards including those for peanut crops, are not as useful as they could be, because they do not accurately describe the essential quality characteristics of the product concerned (1).

At the time the grades and standards for peanut crops were first being formulated, information on peanut quality was grossly lacking. Changes in the marketing grades and standards of peanuts that have since occurred have been due primarily to economic pressures resulting from the dominance of production, our increased awareness of mycotoxins, and buyer demands. Thus, information on peanut quality has been of only minor importance in prompting those changes, especially since the information is widely distributed.

Over the past 15 to 20 years much has been published to enhance our understanding of this difficult-to-define attribute, peanut quality. This paper reviews the peanut quality literature, indicates areas of fruitful research, and suggests potential applications of quality determinations to the marketing process.

Off-Flavor Due to High Temperature Curing

That curing affects the flavor of peanuts was recognized as early as the late 1940's and early 1950's (2,3). Bulk curing temperatures over 35 to 38°C were recognized as having a deleterious effect on quality and flavor (4,5,6). Rapid removal of moisture and curing temperatures above 38°C appeared to be the most important factors associated with loss of quality and impairment of flavor. Subsequently, a generally accepted maximum of 35 to 38°C has been recommended for curing peanuts (7,8,9). Of the various curing temperatures he tested, Pickett (10) found 49°C to be the least satisfactory for flavor and aroma quality. Peanuts bulk-cured in an O_2 atmosphere formed better flavored products than those cured in N_2 or CO_2 atmospheres (11). Whitaker and Dickens (12) reported that the level of off-flavor in peanuts was related to the amount of anaerobic respiration which occurred during curing. They also showed that the off-flavor of immature peanuts cured at 35°C was more intense than that of mature peanuts cured at 52°C. Holaday et al. (13) used a taste panel to evaluate samples of peanut butter prepared from peanuts representing 5 curing methods, 5 temperatures, 3 peanut types, and 3 maturity stages. The flavor of Virginia-type peanuts was significantly affected by curing temperature, whereas that of Spanish-type, by curing method.

Beasley and Dickens (14) observed: (1) Improper curing treatment is more detrimental to the flavor of immature peanuts than to that of mature peanuts. (2) The effect of temperature on flavor produced by high temperature is time dependent. At high temperatures, rapid drying results in less off-flavor than slow drying. At low temperatures, slow drying seems to enhance flavor slightly more than rapid drying. (3) When subjected to high temperature, peanuts at 25% moisture developed more off-flavor than peanuts at other moisture contents. They also found that a water emulsion of off-flavored peanuts had a slightly lower pH than that of peanuts with normal flavor. The skins of peanuts cured at high temperature were light colored and more pink than those cured at low temperatures.

Although several investigators had related high-temperature curing to aroma and flavor defects, Pattee et al. (15,16) were the first to isolate, identify, and compare the volatiles from normal-flavored and off-flavored peanuts. Both qualitative and quantitative differences in the volatile profiles were shown. They suggested that acetaldehyde, ethanol, and ethyl acetate were the volatiles primarily responsible for the off-flavor caused by high-temperature curing. Their suggestion agrees with the hypothesis that off-flavor arises by an anaerobic process. Singleton et al. (17) compared the volatile profiles of peanuts cured at 22°C, 35°C, 45°C, and 50°C. Acetaldehyde, ethanol, and ethyl acetate were again related to flavor deterioration. Increases in their concentrations were detected with each increase in curing temperature. Ethyl acetate was not detected in peanuts cured at 22°C; thus, its presence was proposed as a specific indicator of quality defect. Singleton et al. (17) analyzed the volatiles of peanut samples by gas chromatography (GC) and found that the ratios of GC peak areas for certain volatiles were related to the organoleptic evaluation of the samples. Therefore, they proposed that volatile profile analysis could be used as a basis for objectively evaluating the quality of raw peanuts. However, if it is to be useful in the marketing system for peanuts, the analysis time must be shortened to no more than 20 min.

Normal Raw Peanut Flavor and Other Factors Which Affects Its Volatile Profile

Pattee et al. (18) isolated and identified the volatiles of "normal-flavored" raw peanuts. They found pentane, acetaldehyde, methanol, acetone, ethanol, and hexanal as major components and methyl formate, octane, 2-butanone, and pentanal as minor components. The characteristic aroma and flavor of raw peanuts were suggested to arise from a physical interaction of the components isolated, with hexanal the most significant contributor to this aroma. Brown et al. (19) isolated the aldehydes and ketones from raw peanuts as their 2,4-dinitrophenylhydrazones. Concentrations of hexanal and octanal exceeded their flavor thresholds. The concentration data suggested that in addition to hexanal, octanal and possibly nonanal and 2-nonenal contribute to the "green or beany" flavor of raw peanuts.

Storage is another factor affecting the volatile profile. The volatiles of shelled and unshelled raw peanuts have been shown to change with storage time, reaching a maximum at between 90 to 120 days' storage under simulated warehouse conditions and under controlled environmental conditions of 7°C and 50% R.H. Although a change in quality was not established it was suggested that a relationship existed between the time required for stored peanuts to produce the maximum amount of volatiles (20) and an observed 3 to 4 months' postharvest aging period required for peanuts to develop maximum roasted-flavor potential (21). Because the production of volatiles by raw peanuts is enzymatically mediated in the presence of O_2, the above mentioned data obtained by Pattee et al. (20) on peanut volatiles must be interpreted as indicative of the potential for peanuts to produce increasing amount of volatiles with increasing storage time and not of the level of volatiles within the peanut kernel. Thus, the data suggest the occurrence of a biochemical aging effect during storage. Such an effect would relate volatile production potential with maximum roasted-peanut flavor potential causatively rather than only casually.

Brown et al. (22,23) reported the following compositions in terms of µmoles per 100 g of oil from Spanish-type (S) and runner-type (R) peanuts: total carbonyls, 116 (S) and 62 (R); dicarbonyls, 36 (S) and 30 (R); ketoglycerides 69 (S) and 24 (R); and monocarbonyls, 10 (S) and 8 (R). They suggested that the differences in carbonyl contents were due to differences in linoleate contents between the two peanut types. A longer storage period for the Spanish-type was postulated to allow more autoxidation to occur even though the peanuts were in cold storage.

Qualitatively, differences in cultivar apparently are not reflected by the volatile profiles of peanuts; however, peak ratio analysis has shown several significant quantitative differences (24). When six cultivars from five foreign countries (Argentina, Australia, Israel, South Africa, and Taiwan) were compared with other cultivars grown in the United States, the foreign cultivars averaged higher values for pentane and lower values for methanol and hexanal. A cultivar of known poor quality, 'Pearl', was included in the study, and the GC-peak-area ratios of selected volatiles for that cultivars were compared with the corresponding ratios for the major marketed cultivars in the U. S. All the cultivars used for the comparison had been grown in Virginia. The incidence of significant differences between the ratios for 'Pearl' and the other cultivars was 37%. This finding prompted the authors to suggest that the volatile profiles of new breeding lines of peanuts be compared against a standard profile so that poor quality lines might be eliminated early in the breeding program. Considerable time and resources would therefore be saved in the development of new commercial cultivars.

Blanching (skin removal) methods can have a significant effect on quality. Water-blanched peanuts have the shortest shelf-life, whereas spin-blanched and unblanched peanuts have the longest and similar shelf lives (25). Too high a temperature during blanching has also been shown to result in an increase in some components and pronounced flavor changes (26). Peanuts blanched at 149°C initially had roasted flavor characteristics, but upon storage for period exceeding 30 days, developed undesirable flavor characteristics.

The method of cryogenic vacuum distillation for determining volatile concentration does not lend itself to rapid analyses. A simple, direct gas chromatographic procedure has been developed (27) and employed for analyzing volatile components from raw peanuts. By that method, Brown et al. (28) analyzed five cultivars grown in North Carolina and Virginia and reported that the GC-peak-area ratios of ethanol/methanol and ethanol/total volatiles were significantly correlated to taste panel flavor scores of the peanuts after roasting. The negative correlation between ethanol and roasted flavor is in agreement with the report of Singleton et al. (17) on ethanol content and raw peanut flavor. The direct gas chromatographic method has recently been modified (29) so that samples of different sizes can be analyzed; the sample of volatiles need not be enriched whether obtained at ambient or elevated temperatures; heating is uniform to enhance volatile elution and, thereby, sensitivity; moisture and air are removed to facilitate mass spectral analysis; the analysis is conducted in a closed system to minimize loss of low molecular weight volatiles during elution and, thus, yield a more reliable profile of volatiles.

The results to date suggest that volatile profile analysis can be useful for quality control. However, no methodology rapid enough for this application has yet been developed. No doubt this deficiency will soon be corrected.

Identification of the compounds which form the volatile profile and flavor of raw peanuts will advance studies concerning the mechanics by which these compounds are produced. For example, the findings in that the volatile profile and activity levels of alcohol dehydrogenase and lipoxygenase change during maturation prompted the hypothesis that the volatiles, e.g., acetaldehyde, ethanol, pentane and hexanal, are related to these enzymes and their substrates (30).

Enzymic Origin of Raw Peanut Flavor

The first data postulating a direct enzymic role in the production of the raw peanut volatiles are shown in Figure 1. Shortly after postulating that pentane and hexanal are produced by the action of peanut lipoxygenase within a peanut slurry, Garssen et al. (31) presented direct evidence that pentane is an enzymic product of the soybean lipoxygenase-linoleic acid reaction. Concurrent with this work, St. Angelo et al. (32) published data with the interpretation that hexanal is the only major volatile product of the peanut lipoxygenase-linoleic acid reaction. Subsequently Pattee and co-workers (33) isolated and purified peanut lipoxygenase and demonstrated that its reaction with linoleic acid produced both pentane and hexanal. They further demonstrated that an anaerobic condition was not required for the production of both pentane and hexanal, thus distinguishing their lipoxygenase system from others that had been reported.

The role of lipoxygenase in the production of flavor volatiles from raw peanuts was further investigated by Singleton et al. (34). In model systems containing linoleic acid, purified peanut lipoxygenase was shown to be primarily responsible for the production of flavor volatiles; and the volatile profiles were almost identical to those of peanut homogenates (Figure 2). The optimum pH in the model system, for production of the flavor volatiles was shown to be between 6.5 and 7.0. Pentane but not

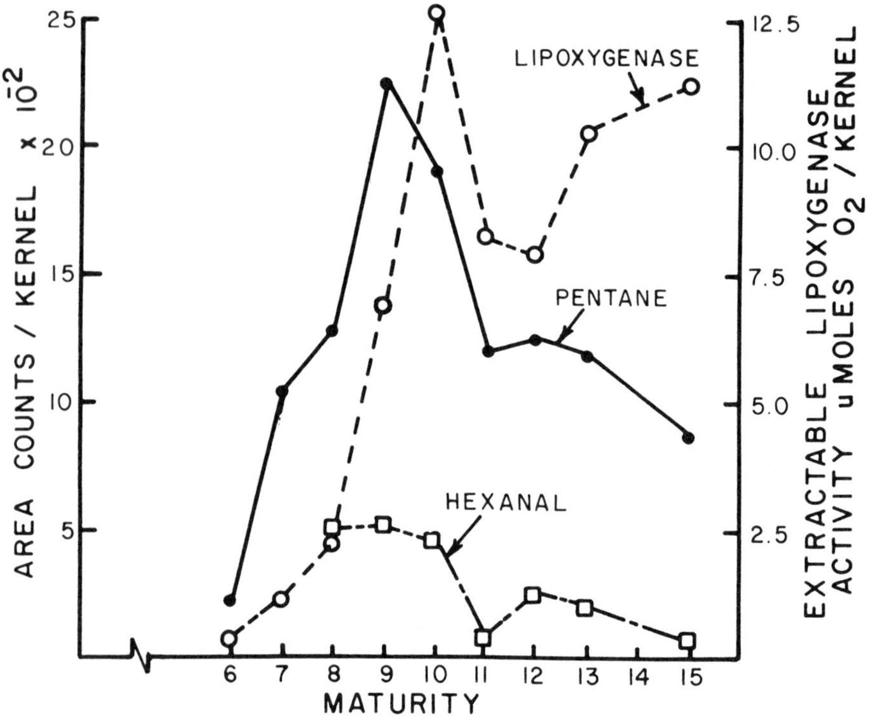

Figure 1. Relationship of lipoxygenase activity and content of pentane and hexanal in the volatile profiles of peanut kernels during maturation (30)

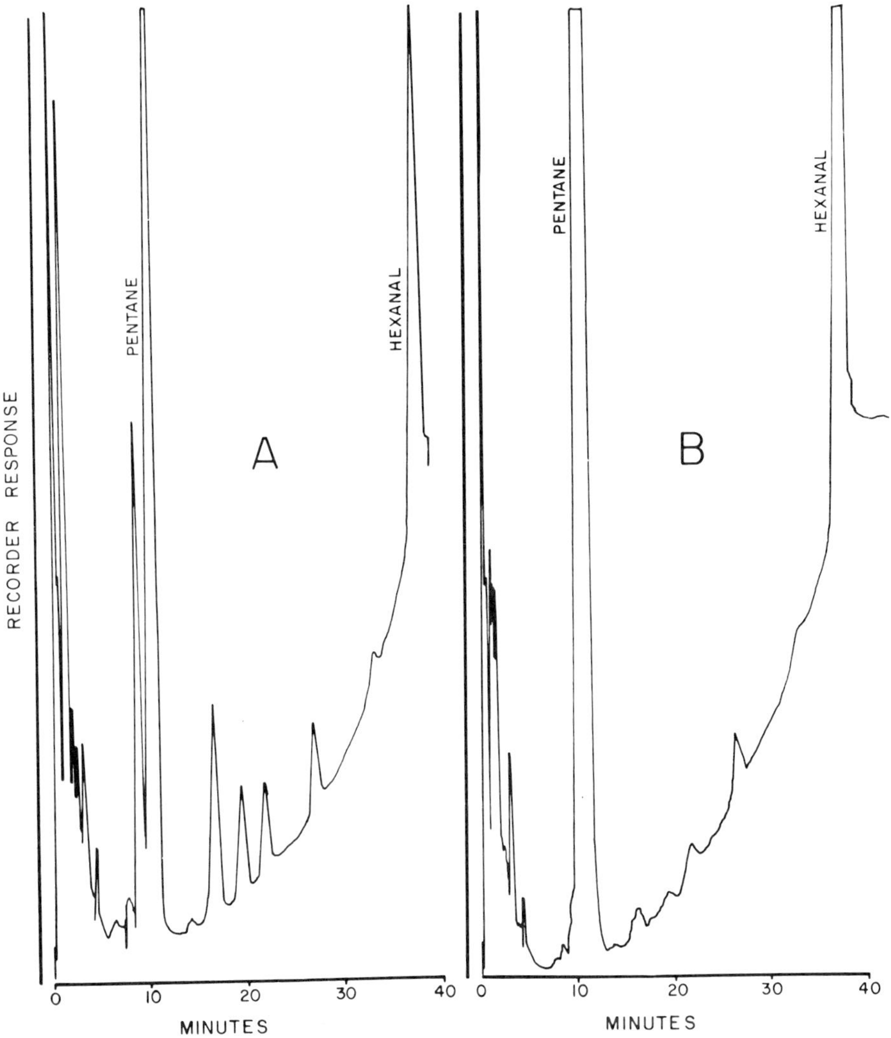

Figure 2. Typical chromatogram of volatiles from a peanut lipoxygenase–linoleic acid model system (A) and a raw peanut homogenate (B) (34)

hexanal production was inhibited by propyl gallate, hydroquinone, and ascorbic acid.

Other factors, such as, pH, O_2 content of the reaction system, and temperature have also been used to differentiate pentane and hexanal production by the peanut lipoxygenase-linoleic system (35). Lowering the pH from the optimum to pH 5.1 reduced the pentane level about 50%, and at pH 4.8 the pentane level was only a trace. The same decreases in pH had no effect on hexanal content. At pH 3.6 pentane and hexanal were found at trace levels. Raising the pH to 8.6 again reduced the pentane level by about 50%, but had little or no effect on hexanal content. At pH 9.6 only a trace of pentane was produced and hexanal production was reduced 50%. Changes in the gaseous atmosphere of the reaction system or the incubation time for the enzyme reaction can also cause changes in the profile. Pentane production was almost completely inhibited when the peanut sample was homogenized in either O_2 or N_2 and the homogenate immediately acidified. Pentane production was partially inhibited when the sample was homogenized in N_2 and the homogenate allowed to stand under aerobic conditions for 15 min. Hexanal was only inhibited when homogenation under N_2 was immediately followed by acidification.

The above results indicate that the lipoxygenase-linoleic acid system produces pentane and hexanal by separate mechanisms from hydroperoxides formed from linoleic acid. Hexanal is produced very rapidly upon formation of the hydroperoxides due to contact of the O_2 atmosphere with the enzyme and substrate. The cucumber flavor volatiles have also been shown to occur with the same rapidity under similar conditions (36,37). Pentane is produced 2-3 minutes after peanuts have been homogenized for for 1-min. Because the pentane reaction is delayed, the sample may be homogenized in an O_2 atmosphere and then acidified to prevent the hydroperoxides from producing pentane. Hexanal production cannot be prevented in this matter.

Recently St. Angelo et al. (38) presented data reaffirming their previous observation that hexanal is an enzymically produced volatile product of the peanut lipoxygenase-linoleic acid reaction. They also confirmed the observations of Pattee and co-workers (33) that pentane is present in the headspace volatiles from the reaction. St. Angelo et al. (38) further showed that care must be excerised to prevent pentane artifacts from arising from the known thermal degradation of hydroperoxide products (39) present in the aqueous solution.

Characteristics of Peanut Lipoxygenase

Siddiqi and Tappel (40) were among the first to demonstrate the presence of lipoxygenase in peanuts. Using crude extracts and assaying them at pH 7.0, they found peanuts to contain only 1% of the lipoxygenase activity found in soybeans. This low activity has perhaps contributed to the lack of investigation of peanut lipoxygenase. Dillard et al. (41), using crude peanut lipoxygenase, found that it showed two pH optima with linoleic acid, trilinolein, and cottonseed oil substrates. This suggested the presence of more than one lipoxygenase isozyme. A pH optimum of about 6 was reported by St. Angelo and Ory (42), who used ammonium sulfate to fractionate the enzyme. No activity was present in the partially purified lipoxygenase preparation at pH 8.0. Several possible reasons would explain the difference in findings of the two groups: the source of linoleate substrate, the method for solubilizing the substrate, and the method for preparing the lipoxygenase. Pattee et al. (33) attempted to isolate a separate pentane-producing enzyme from peanuts, and purified peanut lipoxygenase with ammonium sulfate and column chromatography. Possibly because they assayed the fractions at only one pH and used a relatively steep NaCl gradient on DEAE-Sephadex, they did not observe the isozyme separation reported later by Sanders et al. (43). These researchers reported the separation of three isozymes due to using DEAE-Sephadex A-50 and a linear gradient of NaCl from 0.04 to 0.25 M. Lipoxygenase 1 has a pH optimum at 8.3, whereas lipoxygenase 2 and 3 have a pH optimum at 6.2. All three isozymes had a molecular weight of 76,000 daltons. Peanut lipoxygenase 1 was strongly activated by 0.5 to 1.0 mM Ca^{++}, and the rate of activation was maximum when the ratio of substrate to Ca^{++} was about 2:1 (44). Peanut lipoxygenase 2 and 3 were activated by Ca^{++} but no one Ca^{++} concentration was associated with the optimum level of activity. Calcium differentially activated peanut lipoxygenase, causing the rate of pentane production to increase much more rapidly than the rate of O_2 consumption by the enzyme reaction. At pH 6.2, in the absence of Ca^{++}, the percentage of hydroperoxide isomers produced in the peanut lipoxygenase-linoleic acid reaction were 74.9% 13-hydroperoxy *cis*-9, *trans*-11-octadecadienoic acid (13 LOOH *cis*,*trans*), 2.6% 13-hydroperoxy *trans*-9, *trans*-11-ocatadecadienoic acid (13 LOOH *trans*,*trans*) and 22.5% 9-hydroperoxy 10, 12-octadecadienoic acid (9 LOOH). The presence of 1 mM Ca^{++} at 6.2 did not significantly affect the percentage distribution of the hydroperoxides produced. However, at pH 8.3, the percentage distribution of hydroperoxides produced was 45.2% 13-LOOH *cis*, *trans*, 10.9% 13-LOOH *trans*,*trans*, and 43.9% 9-LOOH in the absence of Ca^{++}, and 57.0% 13-LOOH *cis*,*trans*, 8.9% 13-LOOH *trans*,*trans*, and 35.0% 9-LOOH in the presence of 1 mM Ca^{++}.

The isolation of the geometrical isomers of hydroperoxyoctadecadienoic acid from the peanut lipoxygenase-linoleic acid reaction (45,46) raised the question as to which were enzymically produced and which were autoxidatively produced. Earlier, St. Angelo et al. (32) had isolated 13-LOOH and suggested it to be the only product of the reaction. By taking optical rotation and optical rotatory dispersion measurements of the geometrical isomers, Pattee and Singleton (46) showed that 13-LOOH *cis,trans* and 9-LOOH *trans,cis* were both enzymic products of the reaction while their *trans,trans* counterparts were autoxidation products.

Roasted Peanut Flavor and Factors Which Affect Its Quality

In 1977, 50.7% of the total supply of shelled peanuts and 74.2% of the edible stock in the U. S. were roasted before processing and consumption (47). Roasting of the peanut can thus be considered a major factor in its processing. Some of the earliest work on the volatiles of roasted peanuts was reported by Pickett and Holley (48). Gaseous components identified included derivatives of tetrahydrofuran, ammonia, hydrogen sulfide, and diacetyl. The major gaseous component was carbon dioxide. Pickett and Holley (48) also postulated the involvement of carbohydrates, primarily sucrose, and amino acids in the peanut roasting reactions. Shortly thereafter, Hoffpauir (49) reviewed the literature on peanut composition and its relationship to processing and utilization. Between 1953 and 1964 nothing in the literature added significantly to information available on roasted flavor. However, during the next decade, the work of Mason and co-workers added significantly to the understanding of roasted peanut flavor. They showed that a low-molecular-weight fraction (possibly from the aleurone grains) was the flavor-precursor fraction (50). The amino acids aspartic acid, glutamic acid, glutamine, asparagine, histidine, and phenylalanine were associated with the production of typical roasted flavor (51). The genesis of the roasted flavor components was studied by Koehler et al. (52) and Mason et al. (53). Some 56 compounds were identified as contributors to the aroma of roasted Spanish peanuts (54-58). Walradt et al. (59) in an independent study using steam distillation, reported 187 volatile components. Omitting overlap between the findings of the two groups headed by Mason and Walradt, a total of 223 volatiles have been reported. The short-chain alkyl aldehydes were proposed as the main contributors to the harsh aroma of peanuts immediately after roasting (55) while the pyrazines have a major role in the typical roasted flavor (54, 60).

Ballschmiester and Dereksen (61) reported nine alcohols isolated from a steam distillate of peanuts. Brown et al. (62) resolved extracts from roasted peanuts into various fractions via classical chemical techniques. A number of acids were found, including acetic, propionic, isobuyric, isovaleric, valeric, heptanoic, decanoic, lauric, myristic, phenylacetic, and dihydroxynapthaleneacetic. Carbonyl compounds reported were hexanal, 2,4-decadienal, and 2-oxooctanal. Tentative evidence was presented for aliphatic lactones, which are powerful flavoring agents.

Young and Holley (63) quantified some of the volatiles released during roasting and found that total volatiles tended to increase as quality declined. The increase was in the total carbonyl fraction. Later Young (64) quantitated the volatiles from roasted high-temperature-cured peanuts and reported increases in mercaptans, carbon dioxide, basic compounds, and carbonyls with increasing curing temperature. Total carbonyls appeared to be the best index of declining quality. Milutinovic et al. (65) have reported a relationship between volatiles and flavor scores of peanuts with different roasting treatments. Fore et al. (66, 67) showed a correlation of certain volatile components to flavor score, using various peanut butters stored for different periods. The ratio of methylbutanal to hexanal was the most significantly related to flavor score.

The effect of artificial lighting during marketing on the flavor of peanuts and their volatiles has been studied by Powers et al. (68), while the effect of growing location on flavor has been studied by Cobb and Swaisgood (69). Neither effect was shown to be highly significant. Effect of microwave-, dry-, and oil-roasting peanuts on the flavor of their peanut butters have been investigated by Young et al. (70). Dry-roasting generally was the least preferred and microwave-roasting only slightly more preferable. Roasting versus frying of peanuts has been evaluated on the basis of peanut butter flavor by Metwalli et al. (71). Fried peanuts were preferred to roasted peanuts.

One of the most extensive works conducted on sensory attributes and roasting of peanuts recently demonstrated statistically significant effects of differences among varieties and roasting conditions on flavor (72). The complexity of the relationships demonstrated in this study highlight the challenges which have faced researchers in the past. Three decades of research, have brought us nearly to the threshold of understanding the relationships between the b ochemical changes of physiological development and maturation in the peanut and the quality of the peanut and its products.

Research Areas of the Future in Peanut Quality

In undertaking to discuss future peanut quality research, we consulted many peanut researchers. The areas of research we discuss could only be limited, and the order in which they are presented has no relevance to importance.

Maturity. Although some work has been done in this area we still do not know the relationship of physiological maturity to acceptable peanut quality. Because of the marketing system, which has U. S. No. 1 Virginias as the most immature fraction, physiological maturity may be more important to Virginia-type peanuts than to runner or Spanish types. The indeterminate growing habit of peanuts adds much to the complexity of this challenge. Many of the location and environmental effects may be interrelated to the maturity-quality interaction.

Seed Composition. Although we know the constituents which react within the seed to produce the roasted flavor, research is needed to determine how their relative concentrations affect flavor. The findings of such determinations would be most useful in the establishment of quality guidelines for the marketing channel. Various investigations are currently underway to examine the relation between calcium and roasted flavor, and to develop rapid methods for analyzing flavor volatiles; to develop a rapid method for analyzing arginine; to examine fatty acids and their relationship to quality; and to determine seed-size/roasted-flavor relationships.

Storage. It is anticipated that future regulations by the Food and Drug Administration and the Occupational Safety and Health Administration will have a significant impact on the handling and storage of peanuts in the marketing channels. We therefore need basic studies on the compositional changes which take place during storage and the effects of these changes on quality. The effects of storage duration and condition need to be further investigated. Considerable research is needed into the effects of general storage practices on the quality of farmers' stock peanuts. The presence of foreign materials in this class of peanuts has the potential to cause serious quality damage during storage. The effects of aeration and temperature on quality during storage also need further investigation. Although outside the scope of this review, microorganisms have a potential to cause severe quality damage during storage, particularly in farmers' stock peanuts, and should therefore be studied further. Research into the areas outlined should provide information we need to evaluate the impact of future regulator changes which might be made regarding the handling and marketing of peanuts.

Literature Cited

1. Hinds, M. K.; Kromer, G. W. "Peanuts - Culture and Uses"; American Peanut Research Education Association, Inc.: Yoaku, TX, 1973; Chap. 20.
2. Bailey, W. K.; Futral, J. G.; Pickett, T. A. Georgia Agric. Exp. Sta. Bull. No. 255; 1947.
3. Bailey, W. K.; Pickett, T. A. Georgia Agric. Exp. Sta. Bull. No. 640; 1952.
4. Bailey, W. K.; Pickett, T. A.; Futral, J. G. Peanut J. and Nut World 1954, 33, 17.
5. Bailey, W. K.; Toole, E. H.; Toole, V. K.; Downe, M. J. Proc. Amer. Soc. Hort. Sci. 1958, 71, 422.

6. Butt, J. L.; Kummer, F. A. Agric. Eng. 1951, 32, 27.
7. Dickens, J. W. Peanut J. and Nut World, 1957, 36, 14.
8. Mills, W. T.; Dickens, J. W. N. C. Agri. Exp. Sta. Bull. No. 405, 1958.
9. Turner, W. K.; Suggs, C. W.; Dickens, J. W. Trans. Amer. Soc. Agric. Eng. 1967, 10, 248.
10. Pickett, T. A. Proc. Peanut Res. Conference, Atlanta, GA, 1957; p. 63.
11. Dickens, J. W. N. C. Agric. Exp. Sta. Research and Farming 1957, 16, 5.
12. Whitaker, T. B.; Dickens, J. W. Proc. Third Natl. Peanut Res. Conference, Auburn, AL, 1964; p. 71.
13. Holaday, C. E.; Cecil, S; Bartlett, R. P. Proc. Third Natl. Peanut Res. Conference, Auburn, AL, 1964; p. 91.
14. Beasley, E. O.; Dickens, J. W. N. C. Agric. Exp. Sta. Tech. Bull. No. 155; 1963.
15. Pattee, H. E.; Beasley, E. O.; Singleton, J. A. Proc. Third Natl. Peanut Res. Conference, Auburn, AL, 1964; p. 66.
16. Pattee, H. E.; Beasley, E. O.; Singleton, J. A. J. Food Sci. 1965, 30, 388.
17. Singleton, J. A.; Pattee, H. E.; Johns, E. B. J. Agric. Food Chem. 1971, 19, 30.
18. Pattee, H. E.; Singleton, J. A.; Cobb, W. Y. J. Food Sci. 1969, 34, 625.
19. Brown, D. F.; Senn, V. J.; Dollear, F. G.; Stanley, J. B. J. Amer. Peanut Res. Ed. Assoc. 1971; 3, 208.
20. Pattee, H. E.; Singleton, J. A.; Johns, E. B. J. Agri. Food Chem. 1971, 19, 134.
21. Cecil, S. R. Personal communication, Dept. Food Sci., Univ. of Georgia, Experiment, GA, 1969.
22. Brown, D. F.; Senn, V. J.; Stanley, J. B.; Dollear, F. G. J. Agric. Food Chem. 1972 20, 700.
23. Brown, D. F.; Senn, V. J.; Dollear, F. G. J. Agri. Food Chem. 1973, 21, 463.
24. Pattee, H. E.; Singleton, J. A. J. Agric. Food Chem. 1972, 20, 1119.
25. St. Angelo, A. J.; Kuck, J. C.; Hensarling, T. P.; Ory, R. L. J. Food Processing Preserv. 1977, 1, 249.
26. Pattee, H. E.; Singleton, J. A. J. Amer. Peanut Res. Ed. Assoc. 1971, 3, 189.
27. Brown, D. F.; Dollear, F. G.; Dupuy, H. P. J. Amer. Oil Chemists' Soc. 1972, 49, 81.
28. Brown, M. L.; Wadsworth, J. I.; Dupuy, H. P.; Mozingo, R. W. Peanut Sci. 1977, 4, 54.
29. Legendre, M. G.; Fisher, G. S.; Schuller, W. H.; Dupuy, H. P.; Rayner, E. T. J. Amer. Oil Chemists Soc. 1979, 56, 552.
30. Pattee, H. E.; Singleton, J. A.; Johns, E. B.; Mullin, B. C. J. Agric. Food Chem. 1970, 18, 353.
31. Garssen, G. J.; Vliengenthart, J.F.G.; Boldingh, J. Biochem. J. 1971, 122, 327.

32. St. Angelo, A. J.; Dupuy, H. P.; Ory, R. L. Lipids 1972, 7, 793.
33. Pattee, H. E.; Singleton, J. A.; Johns, E. B. Lipids 1974, 9, 302.
34. Singleton, J. A.; Pattee, H. E.; Sanders, T. H. J. Food Sci. 1976, 41, 148.
35. Singleton, J. A.; Pattee, H. E.; Sanders, T. H. J. Food Sci. 1975, 40, 386.
36. Fleming, H. P.; Cobb, W. Y.; Etchells, J. L.; Bell, T. A. J. Food Sci. 1968, 33, 572.
37. Hatanaka, A.; Kajiwara, T.; Harada, T. Phytochemistry 1975, 14, 2589.
38. St. Angelo, A. J.; Legendre, M. G.; Dupuy, H. P. Lipids 1980, 15, 45.
39. Evans, C. D.; List, G. R.; Hoffman, R. L.; Moser, H. A. J. Amer. Oil Chemists' Soc. 1969, 46, 501.
40. Siddiqi, A. M.; Tappel, A. L. J. Amer. Oil Chemists' Soc. 34, 529.
41. Dillard, M. G.; Henick, A. S.; Koch, R. B. Food Res. 1960. 25, 544.
42. St. Angelo, A. J.; Ory, R. L. "Seed Proteins: Synthesis, Properties, and Processing", Inglett, G. E., Ed.; AVI Publishing Co., Westport, CT, 1972, p. 284.
43. Sanders, T. H.; Pattee, H. E.; Singleton, J. A. Lipids 1975, 10, 681.
44. Nelson, M. S.; Pattee, H. E.; Singleton, J. A. Lipids 1977, 12, 418.
45. Pattee, H. E.; Singleton, J. A. J. Amer. Oil Chemists' Soc. 1977, 54, 183.
46. Pattee, H. E.; Singleton, J. A. J. Agric. Food Chem. 1979, 27, 216.
47. U. S. Department of Agriculture, "Agricultural Statistics 1979" U. S. Government Printing Office, Washington, DC 20402.
48. Pickett, T. A.; Holley, K. T. Georgia Agric. Exp. Sta. Tech. Bull. No. 1; 1952.
49. Hoffpauir, C. L. J. Agric. Food Chem. 1953, 1, 668.
50. Mason, M. E.; Waller, G. R. J. Agric. Food Chem. 1964, 12, 274.
51. Pattee, H. E.; Beasley, E. O.; Singleton, J. A. Proc. Third Natl. Peanut Res. Conference, Auburn, AL; 1964, p. 66.
52. Koehler, P. E.; Mason, M. E.; Newell, J. A. J. Agric. Food Chem. 1969, 17, 393.
53. Mason, M. E.; Newell, J. A.; Johnson, B. R.; Koehler, P. E.; Waller, G. R. J. Agric. Food Chem. 1969, 17, 728.
54. Mason, M. E.; Johnson, B. R.; Hamming, M. C. J. Agric. Food Chem. 1966, 14, 454.
55. Mason, M. E.; Johnson, B. R.; Hamming, M. C. J. Agric. Food Chem. 1967, 15, 66.
56. Johnson, B. R.; Waller, G. R.; Burlingame, A. L. J. Agric. Food Chem. 1971, 19, 1020.

57. Johnson, B. R.; Waller, G. R.; Foltz, R. L. J. Agric. Food Chem. 1971, 19, 1025.
58. Shu, C. K.; Waller, G. R. J. Food Sci. 1971, 36, 579.
59. Walradt, J. P.; Pickett, A. O.; Kinlin, T. E.; Muralidhara, R.; Sanderson, A. J. Agric. Food Chem. 1971, 19, 972.
60. Koehler, P. E.; Mason, M. E.; Odell, G. V. J. Food Sci. 1971, 36, 816.
61. Ballschmieter, H.M.B.; Dereksen, A. W. Fette Siefen Anstrichm. 1970, 72, 719.
62. Brown, B. A.; Konigsbacher, K. S.; Ellison, E. E.; Mann, G. E. J. Food Sci. 1968, 33, 595.
63. Young, C. T.; Holley, K. T. Georgia Agric. Exp. Sta. Bull. No. 41; 1965.
64. Young, C. T. J. Food Sci. 1973, 38, 123.
65. Milutinovic, L.; Bargmann, R. E.; Chang, K. Y.; Chastain, M.; Power, J. J. J. Food Sci. 1970, 35, 224.
66. Fore, S. P.; Goldblatt, L. A.; Dupuy, H. P. J. Amer. Peanut Res. Ed. Assoc. 1972, 4, 177.
67. Fore, S. P.; Dupuy, H. P.; Wadsworth, J. I. Peanut Sci. 1976, 3, 86.
68. Powers, J. J.; Cox, R. J.; Elland, M. C.; Ackroyd-Kelly, P. A. The Flavor Industry 1971, 2, 87.
69. Cobb, W. Y.; Swaisgood, H. E. J. Food Sci. 1971, 36, 538.
70. Young, C. T.; Young, T. G.; Cherry, J. P. Proc. Amer. Peanut Res. Ed. Assoc. 1974, 6, 8.
71. Metwalli, S. M.; Zoueil, M. E.; Mohamed, M. S.; Esmant, M. E. Alex. J. Agric. Res. 1975, 23, 131.
72. Buckholz, Jr., L. L.; Daun, H.; Stier, E.; Trout, R. J. Food Sci. 1980, 45, 547.

RECEIVED April 15, 1981.

14

Instrumental and Sensory Characteristics of Roasted Peanut Flavor Volatiles

LAWRENCE L. BUCKHOLZ, JR.
International Flavors and Fragrances, 1515 Highway 36, Union Beach, NJ 07735

HENRYK DAUN
Rutgers—The State University, Cook College, Department of Food Science, New Brunswick, NJ 08902

Roasted peanuts (Arachis hypogaea) possess a unique and widely enjoyed flavor. In this review we will cover the formation of peanut flavor volatiles, analytical aspects, sensory analysis and correlation of instrumental and sensory results. The roasting process (pyrolysis) subjects the peanut to internal temperatures of 130°C to 150°C (Pickett and Holley) (1), during which the typical roasted peanut flavor is produced. The purpose of pyrolytically treating certain foods promotes flavor changes (common to all of them) that ultimately increase their palatability. Flavor volatiles are produced in these products which influence their aroma and taste characteristics. A knowledge of the composition of these volatiles and the organoleptic changes in their taste characteristics would help us better understand their flavor chemistry.

There are close to 300 flavor compounds identified in roasted peanuts. Walradt et al (2) identified 187 compounds, 142 of them for the first time. Wu (3), listed a table of 205 compounds with references. The most extensive list of compounds was published in 1973 (4). In this review 270 compounds were reported with references. They included 34 hydrocarbons, 26 alcohols, 60 carbonyls, 29 acids, 15 esters, 59 bases, 15 S-compounds and 32 miscellaneous compounds.

Flavor Precursors

The most important constituents of peanuts responsible for the flavor formation during roasting are the amino acids, sugars, proteins and lipids. Typical precursor amino acids are aspartic acid, asparagine-glutamine, glutamic acid, phenylalanine and histidine. Mason et al (5) and Young and Mason (6) noted that the arginine content of Spanish peanut decreased with maturity. Immature nuts contained as much as 50% more arginine than mature nuts. Resultant products (peanut butter) made from immature peanuts were inferior in flavor when compared to peanut butter made from mature nuts.

0097-6156/81/0170-0163$05.00/0

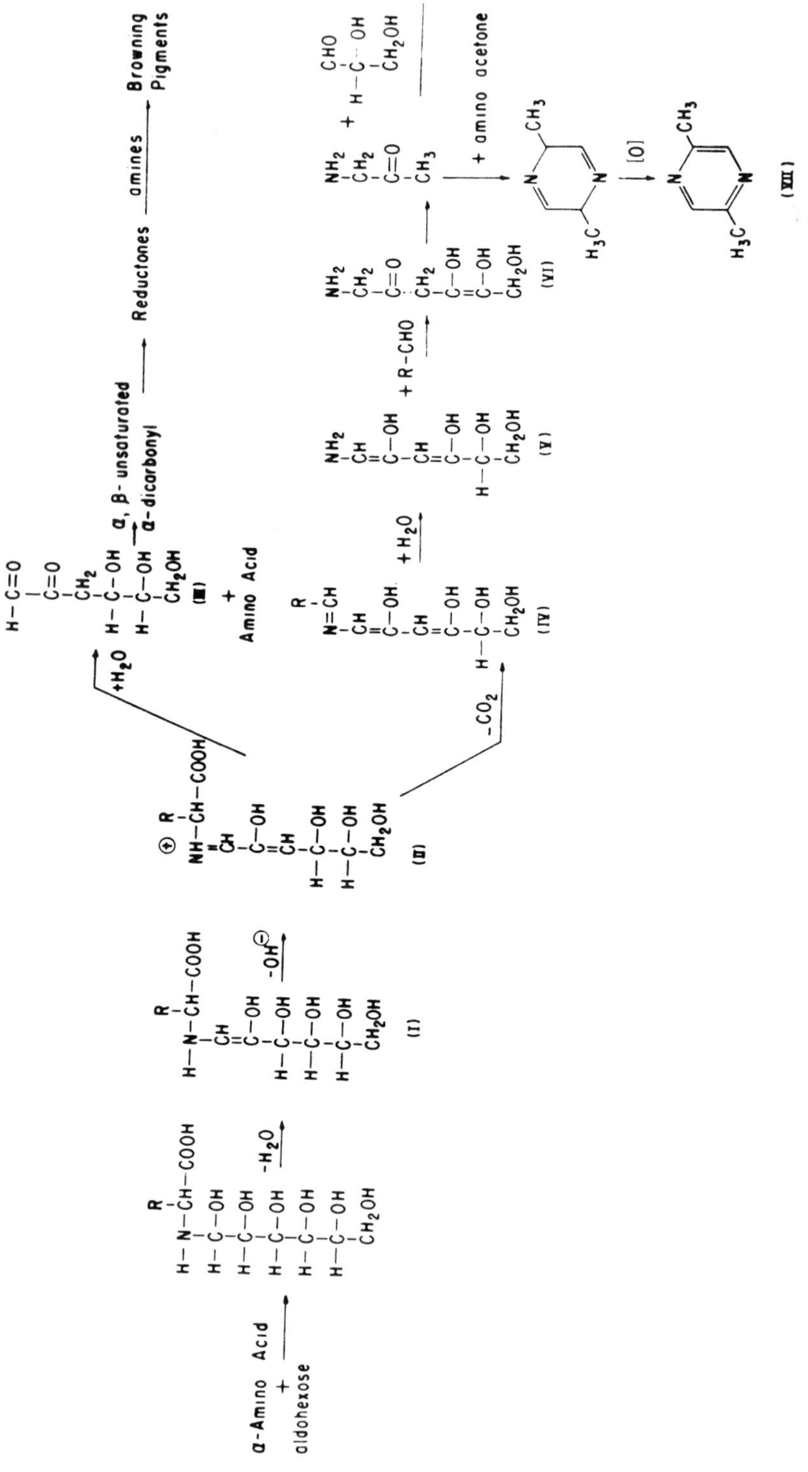

Figure 1. Pyrazine formation mechanism

Products that possess pyrolytically formed flavors such as peanuts, hazelnuts, coffee, cocoa etc. all contain many classes of compounds responsible for the roasted nutty notes.

Investigative work has been done in the development of flavor components from roasting; the low molecules weight carbonyls, the pyrazines and other compounds have all been studied.

Formation of Pyrazines

The alkylated pyrazines formation pathway in amino acid carbohydrate systems was studied by Koehler et al (7).

The effect of roasting on pyrazine formation was demonstrated by Koehler and Odell (8) who used variations of a model system to demonstrate this effect.

As far back as 1967, Newell (9) proposed a pyrazine formation mechanism between sugar and amino acid precursors (see Figure 1). The Schiff base cation is formed by addition of the amino acid to the anomeric portion of aldohexose followed by a loss of water and hydroxyl ion. Decarboxylation forms an imine which can hydrolyze to an aldehyde and dienamine. Enolization occurs and the resulting keto-amine condenses to amino acetone and glyceraldehyde. The pyrazine (2,5 dimethyl) is then formed by the condensation of the two molecules of amino acetone.

Walradt et al (2) investigated the postulated mechanism for the formation of acetyl and methyl acetyl pyrazines by Wange et al (10). Walradt proposed a mechanism for the formation of 6,7-dihdyro-5H cyclopentapyrazine as a result of the interaction of glyoxal and pyruvaldehyde with amino acids and 2-hydroxy-3-methyl-2-cyclopenten-1-one, a frequently occurring product of carbohydrate degradation.

Carbonyls

Various workers have discussed the carbonyl compounds derived from lipids; (11, 12, 13). According to Wu (3), a number of aroma compounds can be traced to lipid degradation during roasting. Aldehydes, ketones, fatty acids, lactones and alcohols were formed. The oxidative degradation of α-amino acids to aldehydes of one less carbon atom by compounds such as alloxan, ninhydrin and 2-furaldehyde was defined by Schonberg (14). He demonstrated that the amino group must be alpha to the carboxyl group and that the carbonyl compound must contain a dione. This reaction which has been known for many years is illustrated in Figure 2. If the hydrogen on the α-carbon of the amino acid is substituted a ketone is produced.

$$RCOCOR' + R''CH(NH_2)COOH \rightarrow R''CHO + CO_2 + RCH(NH_2)COR'$$

Figure 2. Strecker degradation mechanism

Mason et al (15, 5) discussed Strecker degradation as a source of carbonyl formation. He and his colleagues isolated and identified the carbonyls (known to arise from the Strecker Degradation of their corresponding amino acids) from roasted peanuts. These included acetaldehyde, 2 and 3 methyl butanal, isobutyraldehyde and phenylacetaldehyde. He found that all of the corresponding amino acids were shown to be present in the raw peanuts and were destroyed during roasting. Mason showed that phenylacetaldehyde and acetaldehyde are produced in relatively large amounts during roasting; this would be expected from the large amounts of free and peptide bound alanine and phenylalanine present.

Analytical Aspects

Collection of Peanut Flavor Components. Many methods have been used to extract volatiles from pyrolytically processed foods including peanuts; atmospheric distillation of peanut oil, high vacuum distillation (Mason et al) (16), steam distillation, solvent extraction and concentration, etc... All of these techniques have been successfully used to isolate flavor components from peanuts and similar products. Mookherjee et al (17) extracted monocarbonyls from potato chips by use of solvent extraction. Herz and Chang (18) developed a unique apparatus, a falling film still which was extremely versatile in isolating volatiles from various food components. Walradt et al (2) used a turbo film evaporator, distilling volatiles from an aqueous ground peanut slurry and collecting the condensate in a series of traps.

Headspace Technique. One of the most recent methods of flavor analysis which evolved with the development of sensitive gas chromatographic instrumentation is the headspace technique. Withycombe et al (19), gave an excellent description of headspace analysis. In this procedure volatiles in gaseous state that are in equilibrium over the food are analyzed.

Brown et al (20) examined volatiles from roasted peanuts by introducing glass liners containing ground samples of roasted nuts into the injection port of a gas chromatograph where volatiles were vaporized in situ.

In many instances concentration of the substances in the headspace is necessary.

Adsorption polymers have been used for collection, concentration, and subsequent G.C. analyses in a wide variety of applications in recent years. Withycombe et al (19) used several adsorption polymers to trap the headspace volatiles from hydrolyzed vegetable protein (HVP) and found that of three polymers investigated, Chromosorb 105, Porapak Q, and Tenax GC, the volatiles trapped on Tenax GC contained the most characteristic HVP aroma.

A method described by Buckholz et al (21) employs the use of the adsorption polymer Tenax GC (para 2,6 diphenylene oxide) to collect volatiles from roasted peanuts. Two types of peanuts Runner #1 and Spanish were roasted.

The temperature was held at a constant 163°C and roasting times of 7 minutes, 8 minutes and 9 minutes were used to obtain light roast, medium roast and dark roast peanuts.

Optimal conditions for the adsorption polymer method collection of volatiles were as follows. Four hundred grams of peanuts were placed in a 50°C jacketed glass column and extracted with nitrogen for 4 hours at a flow rate of 40 mls/min. Volatiles were adsorbed onto eighth inch traps packed with Tenax GC which were attached to the top of the column via a teflon thermometer adapter. Three traps were used in series. Six collections were made for each roasting condition.

Gas Chromatography - Mass Spectrometry. The development of gas chromatography greatly increased the analytical chemists ability to separate, isolate and identify components. Gas chromatography (GC) has been used in conjunction with mass spectrometry, nuclear magnetic resonance spectrometry and infra red spectroscopy as evidenced by Mason et al (16) in his peanut flavor work. Bondarovich et al (22) used mass spectrometry in conjunction with GC to identify pyrazines in coffee. van Praag et al (23) used similar instrumental techniques on cocoa which is also high in pyrazines. Pattee et al (24) analyzed raw peanut essence by GC mass spectral means. Raw peanut essence is high in carbonyls. Walradt et al (2) used a 500 ft. stainless steel capillary column .03 inches in diameter coated with Carbowax 20M coupled to a mass spectrometer to identify the volatiles in his peanut essence. Johnson et al (25, 26) used high and low resolution mass spectrometers coupled to open tubular columns and Crain and Tang (27) identified components in Macadamia nuts by GC mass spectrometry and compared their findings to GC retention indices and mass spectral reference data. So GC mass spectral analysis has proved to be an excellent tool for the separation, identification of peanut volatiles.

In a recent work, Buckholz et al (21), the traps containing adsorbed volatiles were inserted directly into the modified injection port of a Varian 2700 gas chromatograph (GC) for analysis. A 400 ft. by 0.032 inch glass capillary column coated with SE-30 was used. The column was temperature programmed from 50°C to 190°C at 2°/min.

Mass spectral analysis was used to identify the GC peaks present in the peanut headspace volatiles.

Seven new compounds were tentatively identified by GC mass spectral analysis.

The newly identified compounds were n-ethyl pyrrole, 1,2-dimethyl pyrrole, 1-octen-3-ol, 2,4-dimethyl-3-thiazoline, ethyl decanoate, decane and indane.

Effect of Roasting Time on the Amount and Composition of Volatiles. Buckholz et al (28) studied the influence of roasting time on the chemical composition of the aroma of fresh roasted peanuts. Analytical details of this work will be published in the near future.

Peaks were quantitated by a chromatography computer using an Internal Standard method (Buckholz et al) (21). Ethyl nonanoate was used as the Internal Standard.

The amount of peanut headspace volatiles adsorbed onto the traps was expressed as ethyl nonanoate.

Table I shows the total amounts of volatiles collected for all roasting times and both peanut types.

Quantities are expressed in 10^{-6} gms. The G.C. profiles were separated into 3 boiling pt. zones; Most Volatile (pks 1-10); Medium Volatile (pks 11-19) and Least Volatile (pks 20-32).

Figure 3 shows GC profile for Light, Medium and Dark roasted samples for Runner #1. The same peaks were found to be present in both peanut types (Runner #1 and Spanish) and under all roasting conditions, but differed quantitatively.

Thirty-two peaks were selected for evaluation based on an initial composition of 0.1% or more.

The carbonyls are represented by peaks 4, 6, 7, 11 and 16 on Figure 3. They were identified as isobutyraldehyde, isovaleraldehyde, 2-methyl butanal, pentanal and hexanal. The pyrazines are represented by peaks 13, 18, 20, 25 and 26. They were identified as pyrazine, 2-methyl pyrazine, 2,5-dimethyl pyrazine, 2-ethyl-5-methyl pyrazine and 2-ethyl-3-methyl pyrazine. The authors observed a decrease in carbonyls (which are responsible for harsh green flavor notes) and increase in pyrazines (which are responsible for roasted notes) with longer roasting time.

2-Methyl butanal is a typical carbonyl which has been previously identified in peanuts, Mason et al (15). 2,5-Dimethyl pyrazine is also a typical compound which has been previously reported in peanuts (Mason et al) (16). A decrease in aldehyde and subsequent increase in pyrazine with increased roasting time was demonstrated by these authors. This supports Mason and Johnson's findings (15, 16). They reported that the low molecular weight carbonyls, particularly aldehydes, were responsible for the harsh green notes present in roasted peanuts while the alkylated pyrazines were responsible for the roasted nutty character of roasted peanuts (15, 16).

Sensory Attributes of Roasted Peanuts

Few foods have a flavor and texture that is as universally liked as peanuts. Typical peanut flavor, nuttiness, sweetness and bitterness can be altered by variety, growing conditions, methods of harvesting, storing and processing (roasting).

The bitter flavor in peanuts is due to at least four components of saponins which are about 20 times as concentrated in

TABLE I

COMPARISON OF THE TOTAL VOLATILES ADSORBED UNDER STANDARD CONDITIONS FOR VARIOUS ROASTING TIMES IN THE MOST (I), MIDDLE (II) AND LEAST (III) VOLATILE AREAS OF THE GC CHROMATOGRAM

CHROMATOGRAM SECTIONS	RUNNER*			SPANISH*		
	LIGHT	MEDIUM	DARK	LIGHT	MEDIUM	DARK
MOST VOL. 1-10	173.63	166.28	153.59	171.80	149.91	145.90
MED. VOL. 11-19	117.65	99.46	124.94	126.29	161.26	161.41
LEAST VOL. 20-32	33.39	50.83	74.21	39.94	60.68	68.95
TOTAL	324.67	316.57	352.74	338.04	371.85	366.26

* - This represents 6 collections and subsequent GC analysis per roasting condition. Quantities expressed in 10^{-6} gms.

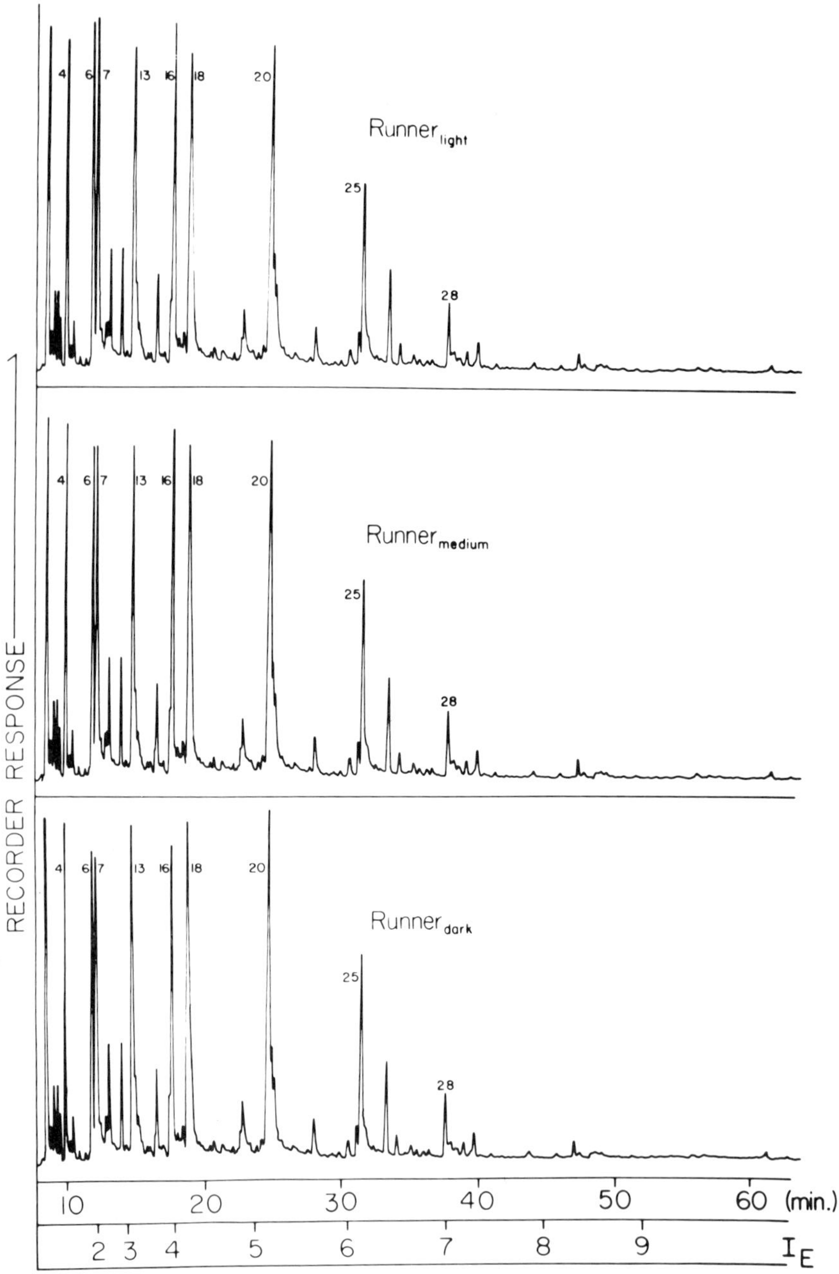

Figure 3. GC profiles for light, medium, and dark roasted samples for Runner #1

the hearts of peanuts as in the cotyledons according to Fisher (29) and Diechert and Morris (30).

Peanut flavor is closely related to the oil, and on separation the flavor goes with the oil rather than with the meal. In practically all cases of heating the flavor is accentuated. This is particularly true with dry roasting.

The pyrazines were responsible for the roasted nutty flavor of fresh roasted peanuts (Mason et al) (15, 16), (Johnson 25, 26). According to Mason et al (15), the low molecular weight aldehydes are responsible for the harsh aroma associated with fresh roasted nuts.

Sensory evaluation is indispensable for evaluating food and flavor materials. Peanuts and peanut products have been extensively evaluated using sensory analysis alone and in combination with instrumental analysis; (Milutinovic et al) (31), (Fore et al) (32), (Dupuy et al) (33, 34,) and (Young et al) (35). Dravnieks et al (32) evaluated corn using sensory evaluation in combination with G.C. Milutinovic used a 9 pt. hedonic scale to rate and determine significant differences among tomato juice blends and peanut samples from various roasting times. The sensory scores were then correlated with instrumental findings. Fore et al (32) rated peanut butter samples using hedonic scaling and correlated these results to gas chromatographic (G.C.) results. G.C. volatile profiles were determined for 14 peanut butters which had been flavor scored. When selected peak area ratios were plotted against flavor scores 9 of the 14 points were almost on the regression line, the rest were within one unit. Dupuy also used hedonic scaling on vegetable oils and various peanut butters and Young et al (35) evaluated peanut butters from roasted peanuts prepared by different roasting methods using a nine point hedonic scale. Triangle testing can be used prior to hedonic rating, to eliminate psychological influences from affecting judges scoring.

The influence of roasting time on sensory attributes of roasted peanuts was examined by Buckholz et al (36). Sensory methods mentioned below are described in detail in this publication.

Peanut samples from each roasting condition were evaluated organoleptically for strength and desirability of aroma and flavor using a 9 point hedonic scale. Statistical analysis was then done using the Tukey Test to determine significant differences among roasting conditions.

Table 2 shows the relative rankings of all four flavor attributes namely strength and desirability of aroma and flavor as influenced by roasting time. This table illustrates the average flavor score for each attribute. The optimum samples chosen on the basis of the mean value for desirability of aroma and flavor were Runner #1 Medium and Spanish Dark.

The Tukey Test of significance at .05 level showed significant differences in roasting conditions with respect to strength of odor and flavor but no significant differences with respect to

TABLE II

RELATIVE RANKING OF SAMPLES USING AVERAGE SCORE

	STRENGTH OF		DESIRABILITY OF	
	ODOR	FLAVOR	ODOR	FLAVOR
RUNNER #1				
LIGHT	4.98	4.78	4.83	4.96
MEDIUM	6.07	5.69	5.32	5.45
DARK	6.84	6.74	4.94	5.11
SPANISH				
LIGHT	5.46	5.26	5.24	5.30
MEDIUM	6.01	5.80	5.28	5.10
DARK	6.21	6.02	5.56	5.19
TUKEY (0.05) Q =	0.81	0.85	0.89	0.93

desirability of odor and flavor Table 3. Details on the Tukey Test as a method of statistical analysis can be found in (Steel et al) (37).

The Tukey Test is a rather conservative statistical calculation that demonstrates only severe differences and ignores subtle ones that might be considered significant by other means of calculation. In this method as shown in Table 3 any two samples that are not directly connected by a bar are declared significantly different from each other.

Contribution of Newly Identified Compounds to Peanut Flavor

To determine the importance of the newly identified compounds in their overall contribution to peanut flavor, the following sensory evaluation was conducted by six experienced flavor chemists in a lab specifically designed for sensory analysis.

A good quality typical peanut flavor was evaluated in distilled water at 50 ppm. This level of flavor provided optimal aroma and taste characteristics. The new compounds were then added individually to the peanut flavor in solution and evaluated organoleptically for their overall contribution to the aroma and flavor of the peanut flavor. The peanut flavor containing the new compound was then evaluated against the peanut flavor alone to determine enhancement or detraction from the existing flavor. Concentrations of the added chemicals were adjusted until optimal organoleptic contribution was perceived. Table 4 shows the peanut flavor concentration in distilled water plus a list of the compounds, optimal concentration in the diluted peanut flavor and a description of the organoleptic contribution or detraction from the flavor.

It was found that the chemicals individually enhanced selected areas of the overall peanut flavor. The n-ethyl pyrrole and ethyl decanoate enhanced the nut meat area of the flavor which contributed to the overall sweet nuttiness. The ethyl decanoate also added sweet fatty notes which contributed to the overall flavor body.

The 2,4-Dimethyl-3-thiazoline added deep roasted notes and contributed to the nut skin character. Some people describe these notes as the shelly notes.

The 1-Octen-3-ol added to the earthy green notes in the background, some fattiness and also enhanced the sweet nut skin portion.

Decane - the hydrocarbon contributed to the fullness or body of the flavor adding some fatty or waxy notes and rounding off the flavor in general which imparted a fullness or richness to the overall flavor character.

Indane - added some harsh solventy notes contributing to the green area and contributed to the fullness of the flavor. 1,2-Dimethyl pyrrole was not evaluated since a reference sample was not available.

TABLE III

TUKEY TEST FOR SIGNIFICANT DIFFERENCES BETWEEN SAMPLES

STRENGTH ODOR

R-L*	S-L*	S-M*	R-M*	S-D*	R-D*
4.98	5.46	6.01	6.07	6.21	6.84

STRENGTH FLAVOR

R-L	S-L	R-M	S-M	S-D	R-D
4.78	5.26	5.69	5.80	6.02	6.74

DESIRABILITY ODOR

R-L	R-D	S-L	S-M	R-M	S-D
4.83	4.93	5.24	5.28	5.31	5.56

DESIRABILITY FLAVOR

R-L	S-M	R-D	S-D	S-L	R-M
4.96	5.10	5.11	5.19	5.30	5.44

* R = RUNNER

S = SPANISH

L = LIGHT

M = MEDIUM

D = DARK

TABLE IV

EVALUATION OF NEWLY IDENTIFIED FLAVOR CHEMICALS FOR THEIR OVERALL CONTRIBUTION TO PEANUT FLAVOR

STANDARD TEST SOLUTION - PEANUT FLAVOR 50 PPM* IN DISTILLED WATER

CHEMICAL	OPTIMUM LEVEL PPM	
N-ETHYL PYRROLE	1.0	ADDS TO SWEET MEATINESS OF PEANUT AND ENHANCES OVERALL NUTTINESS.
ETHYL DECANOATE	.01	ADDS SWEET FATTY NOTE WHICH CONTRIBUTE TO OVERALL FULLNESS OF FLAVOR
2,4-DIMETHYL-3-THIAZOLINE	.05	ADDS TO PEANUT SHELL CHARACTER ENHANCES ROASTED NUT SKIN NOTES AND ADDS DEPTH TO THE ROASTED NOTES.
1-OCTEN-3-OL	.01	ADDS EARTHY SWEET NUTTINESS. ALSO ADDS FULLNESS TO FLAVOR CHARACTER.
DECANE	2.0	CONTRIBUTES A UNIQUE FATTINESS ALMOST A WAXY NOTE - ROUNDING OFF FLAVOR AND ENRICHING THE FLAVOR BODY.
INDANE	0.1	ADDED SOME HARSH SOLVENTY NOTES CONTRIBUTING TO THE GREEN AREA AND CONTRIBUTED TO THE FULLNESS OF FLAVOR.

* - This was found to be the optimum level for this flavor in distilled water.

More and more work is being done in correlating instrumental analysis with sensory evaluation. McCarthy et al (38) and Heinz et al (39) did earlier correlation studies on fruit. More recent works were done by Dupuy et al (40) and Warner et al (41). They correlated sensory score vs G.C. analysis on vegetable oils obtaining excellent results.

Powers (42) provided a general description of a computer assisted method for correlating volatiles from foods with sensory impressions using the gas chromatograph and organoleptic data. First, profile curves are treated to yield peak heights, peak areas or ratios between peaks based on area or height. A computer program then picks out these variables which are the most discriminating factors between profile samples. These factors are then mathematically related back to the organoleptic scores of the samples from which the profile curves were prepared. Weighting factors for each peak variable are then calculated. When the discriminating peak variables and weightings are known, a discriminant equation can be used to predict the organoleptic scores of samples which are related to, but not necessarily part of the initial sample set examined. The larger the initial set of samples from which profiles are available, the closer the correlation between predicted organoleptic scores and peak variables becomes. This correlation is strictly mathematical and has no relation to flavor compounds affecting organoleptic scores.

This technique has been modified by other authors. Biggers et al (43) on coffee, Milutinovic et al (31) on peanuts and tomato products, Powers et al (44) on peanuts and Dravnieks et al (45) on corn.

Blumenthal et al (46) correlated G.C. chromatograms with organoleptic scores on various fats and oils after simulated deep fat frying. A computerized statistical approach showed excellent correlation between G.C. peaks and organoleptic scores.

Correlation of Instrumental and Sensory Data

Buckholz et al (36) selected thirty-two peaks from the GC profiles to be quantitated and correlated to organoleptic flavor scores. The chosen peaks were determined to be important to peanut flavor and common to all profiles for both types of peanuts and all three roasting conditions. Their averaged individual areas were entered onto punch cards along with organoleptic panel scores for the four flavor attributes for all peanut roast samples. Stepwise multiple regression was then used to correlate instrumental quantitation to flavor score.

Figures 4 and 5 show individual peaks correlated to flavor score.

Peak 11 was identified as pentanal. This was a typical carbonyl which had been identified in peanuts. Organoleptically this compound represented harsh, green solventy notes. The negative correlation -0.9770 (Coefficient of Determination) showed

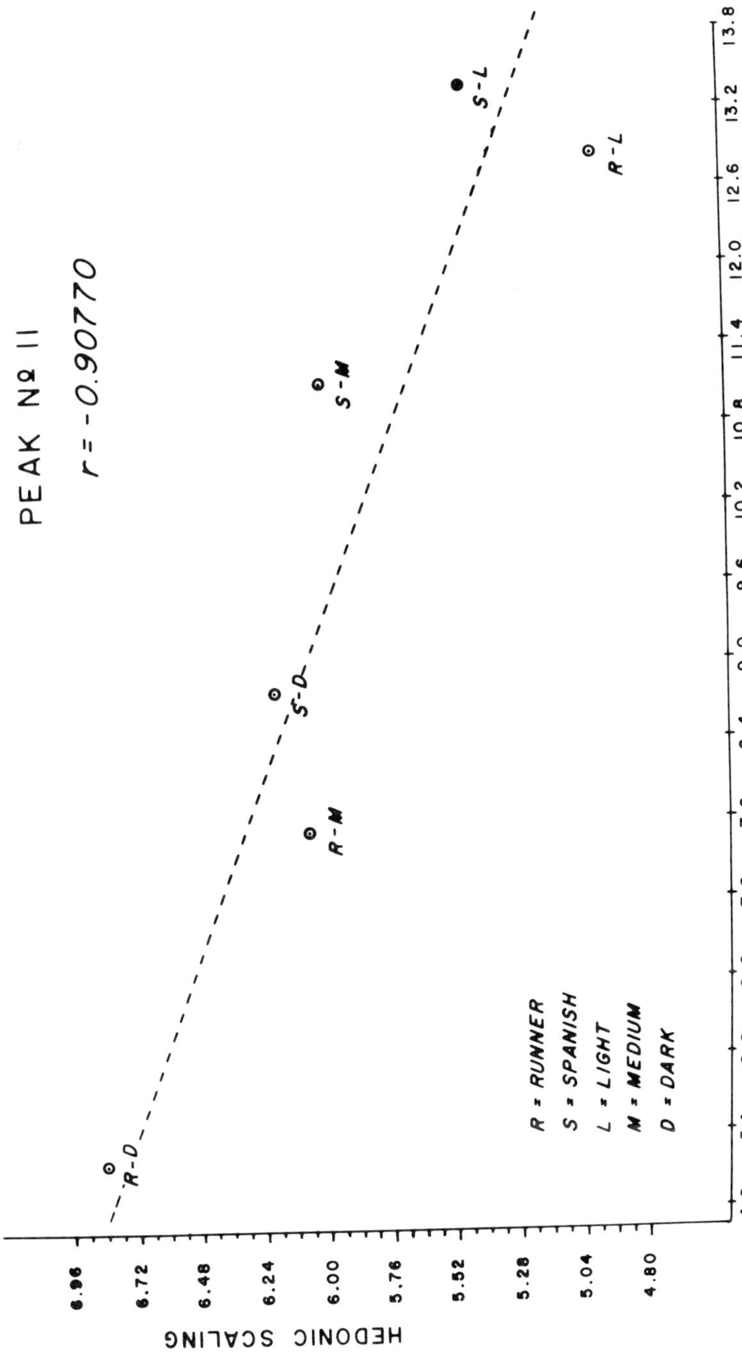

Figure 4. Correlation of peak 11 (pentanal) to flavor score

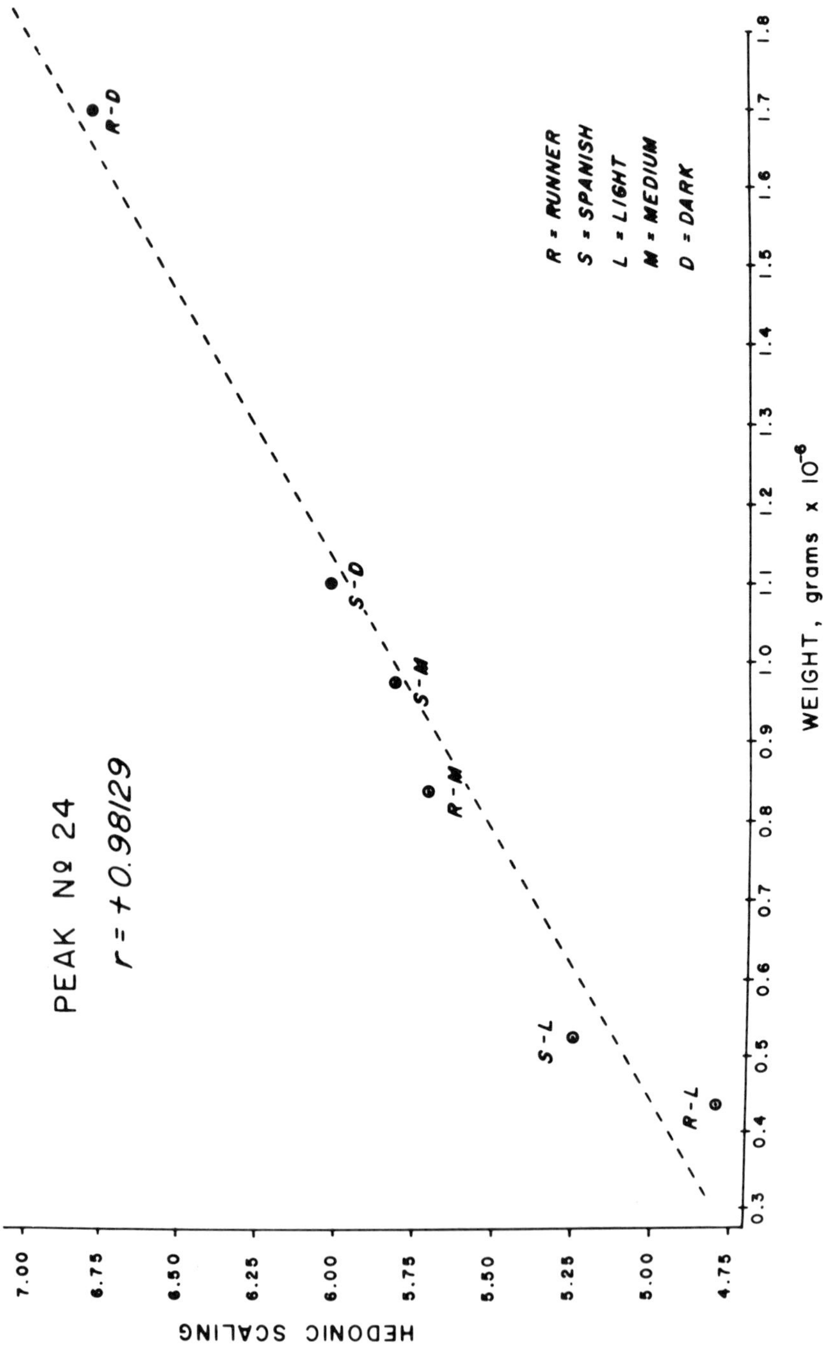

Figure 5. Correlation of peak 24 (2-ethyl-6-methyl pyrazine) to flavor score

that sensory preference increased as concentration decreased. Peak 24 was identified as 2-ethyl-6-methyl pyrazine. This was a typical pyrazine which had been identified in peanuts. Organoleptically this compound represented roasted, nutty, earthy notes. The positive correlation of +0.98129 (C.D.) showed that sensory preference increased with concentration. These plots indicated that a decrease in carbonyls with a subsequent increase in pyrazines were important to good quality peanut flavor. This supported the work of Mason et al (15, 16) and Johnson (25, 26).

Conclusions

The aroma of fresh roasted peanuts is influenced by the roasting time and is a reflection of the changes in the ratios of carbonyl derivatives to pyrazines.

A knowledge of the mechanism of formation of these volatiles is therefore not only desirable, but necessary to optimize quality and overall flavor acceptance; or where duplication and/or modification of the particular flavor is desired.

By changing roasting time customized flavors can be created to provide the desired type of peanut quality.

Peanuts roasted for various lengths of time can be successfully rated by hedonic scaling if the evaluation is separated into four sensory attributes; namely strength and desirability of aroma and flavor. Panelists did readily agree on the intensity of aroma and flavor but not desirability.

Correlation of instrumental and sensory data showed that a decrease in carbonyls with a subsequent increase in pyrazines is important for good quality peanut flavor.

It was demonstrated that the newly identified compounds all had a significant influence on the aroma and flavor of fresh roasted peanuts. Evaluation in an existing flavor showed that all the compounds with the exception of indane produced a marked increase in the quality of the peanut flavor.

Abstract

The roasting process subjects peanuts to internal temperatures of 130°-150°C during which the typical roast peanut flavor is produced. The most important constituents of peanuts responsible for flavor formation are the amino acids, sugars, protein and lipids. There are close to 300 flavor compounds identified in roasted peanuts. Pyrazines contribute roasted nutty aroma and aldehydes are responsible for harsh green aroma. Recent development in the research include application of an absorption polymer Tenax GC to collect volatiles, quantitation of flavor components and computer aided correlation of instrumental and sensory data.

Literature Cited

1. Pickett, T. A.; Holley, K. T. 1952, Ga. Agr. Exp. Sta. Tech. Bull. 1.
2. Walradt, J. P.; Pittet, A. O.; Kinlin, T.E.; Muraldihara, R.; Sanderson, A. J. Agr. Food Chem. 1971, 19, 972.
3. Wu, C. The Isolation and Fractionation of Volatile Flavor Constituents of Roasted Peanuts, Ph.D. Thesis, Rutgers University, New Brunswick, New Jersey.
4. van Straten, S.; de Vrijer, F., (Eds.), "Lists of Volatile Compounds in Food", 3rd Edition, Central Institute for Nutrition and Food Research, Zeist, Netherlands, 1973.
5. Mason, M. E.; Newell, J.A.; Johnson, B.R.; Koehler, P.E.; Waller, G. R. J. Agr. Food Chem. 1969, 17, 728.
6. Young, C. T.; Mason, M. E. J. Food Sci. 1972, 37, 722.
7. Koehler, P. E.; Mason, M. E.; Newell, J. A. J. Agr. Food Chem. 1969, 17, 393.
8. Koehler, P. E.; Odell, M. E. J. Agr. Food Chem. 1970, 18, 895.
9. Newell, J. A.; Mason, M. E.; Matlock, R.S. J. Agr. Food Chem. 1967, 15, 767.
10. Wange, P. S.; Kato, H.; Fujimaki, M. Agr. Biol. Chem. (Japan) 1969, 33, 1775.
11. Forss, D. A. Progr. Chem. Fats Other Lipids 1972, 13, 177.
12. Dismick, P. S. "The Formation and Characteristics of Flavor Components Derived from Food Lipids", 29th P.M.C.A. Production Conference, 1975; p 25-31.
13. Erikkson, C. J. Agr. Food Chem. 1975, 23, 126.
14. Schonberg, A.; Moubacker, R.; Mostofa, A. J. Chem. Soc. 1948, 176.
15. Mason, M. E.; Johnson, B. R.; Hamming, M. C. J. Agr. Food Chem. 1967, 15, 66.
16. Mason, M. E.; Johnson, B. R.; Hamming, M. C. J. Agr. Food Chem. 1966, 14, 454.
17. Mookherjee, B. D.; Deck, R. E.; Chang, S. S. J. Agr. Food Chem. 1965, 13, 131.
18. Herz, K. O.; Chang, S. S. J. Food Sci. 1967, 31, 937.
19. Withycombe, D. A.; Mookherjee, B. D.; Hruza, A. "Analysis of Foods and Beverages Headspace Techniques", Academic Press, New York, 1978; p 81.
20. Brown, D. F.; Dollear, F. G.; Dupuy, H. P.; J. Am. Oil Chem. Soc. 1972, 49(1), 81.
21. Buckholz, L. L., Jr.; Withycombe, D. A.; Daun, H. J. Agr. Food Chem. 1980A, 28, 760.
22. Bondarovich, H. A.; Friedel, P.; Krompl, V.; Renner, J. A.; Shepherd, F. W.; Gianturco, M. A. J. Agr. Food Chem. 1967, 15(6), 1093.
23. van Praag, M.; Stein, H. S., and Tibbetts, M.S. J. Agr. Food Chem. 1968, 16, 1005.

24. Pattee, H. E.; Singleton, J. A.; Cobb, W. Y. J. Food Sci. 1969, 34, 625.
25. Johnson, B. R.; Waller, G. R.; Burlingance, A. L. J. Ag. Food Chem. 1971A, 19, 1025.
26. Johnson, B. R.; Waller, G. R.; Foltz, R. L. J. Agr. Food Chem. 1971B, 19, 1025.
27. Crain, W. O., Jr.; Rang, C. S. J. Food Sci. 1975, 9, 541.
28. Buckholz, L. L., Jr.; Hruza, A.; Daun, H.; Chang, S., presented at the 13th Middle Atlantic Regional Meeting of the American Chemical Society, Monmouth College, West Long Branch, N.J., March 19-23, 1979.
29. Fisher, C. H. Peanut J. Nut World 1959, 38(6), 14.
30. Diechert, J. W.; Morris, N.J. J. Agr. Food Chem. 1958, 6 930.
31. Milutinovic, L.; Borgmann, R. E.; Chang, K. Y.; Chatain, M.; Powers, J. J. J. Food Sci. 1970, 35, 224.
32. Fore, S. P.; Goldblatt, L. A.; Dupuy, H. P. J. Amer. Peanut Res. Educ. Assoc. 1972, 4, 1.
33. Dupuy, H. P.; Fore, S. P.; Goldblatt, L. A. J. Am. Oil Chem. Soc. 1973, 51, 140.
34. Dupuy, H. P.; Fore, S. P.; Wadsworth, J. I. J. Am. Oil Chem. Soc. 1974, 51, 140.
35. Young, C. T.; Young, T. G.; Cherry, J. P. J. Amer. Peanut Res. Educ. Assoc. 1976, 4(1), 8.
36. Buckholz, L. L., Jr.; Daun, H.; Stier, E.; Trout, R. J. Food Sci. 1980B, 45(3), 547.
37. Steel, R. G. D.; Torrie, J. H. "Principles and Procedures of Statistics", McGraw Hill, New York, Toronto, London, 1960; p. 110.
38. McCarthy, A. I.; Polmer, J. K.; Shaw, C. P.; Anderson, E. F. J. Food Sci. 1962, 27, 379.
39. Heinz, D. E.; Pangborn, R. M., Jennings, W. C. J. Food Sci. 1964, 29, 756.
40. Dupuy, H. P.; Rayner, E. T.; Wadsworth, J. I. J. Am. Oil Chem. Soc. 1976, 53, 628.
41. Warner, K.; Evans, C. D.; List, G. R.; Dupuy, H. P.; Wadsworth, J. I.; Goheen, G. E. J. Am. Oil Chem. Soc. 1978, 55, 252.
42. Powers, J. J. Food Technol. 1968, 22, 383.
43. Biggers, R. E.; Hilton, J. J.; Gianturco, M. A. J. Chromatogr. Sci. 1969, 7, 453.
44. Powers, J. J.; Cox, R. J.; Ellanel, M. C.; Ackroyal-Kelly, P. A. J. Agr. Food Chem. 1971, 22(5), 744.
45. Dravnieks, A.; Reilick, H. G.; Whitfield, J.; Watson, C. A. J. Food Sci. 1973, 38, 34.
46. Blumenthal, M. M.; Trout, J. R.; Chang, S. S. J. Am. Oil Chem. Soc. 1976, 53, 496.

RECEIVED April 7, 1981.

15

Comparison of the Food Value of Immature, Mature, and Germinated Soybeans

J. J. RACKIS

Northern Regional Research Center, Agricultural Research, Science and Education Administration, U.S. Department of Agriculture, Peoria, IL 61604

Wolf (1) has pointed out that the soybean could be classified as a protein seed rather than an oilseed. Average proximate composition of presently grown varieties is 40% protein and 20% oil; depending upon variety and strain, protein content ranges between 34-49% and oil content ranges from 14-25% (2). Protein content in soybeans is higher than in many meat products and about twice that of other food legumes on an as-is basis.

The soybean crop, particularly in the United States, is a big protein reserve used mainly for conversion to animal protein and for export. Production of edible soy protein products, estimated to be 100 million pounds in 1967, was increased to about one billion pounds in 1977 (3). These soy protein products are reasonably well-tolerated and are of good protein value for humans of all ages (4,5,6). Indications are that considerable amounts of more expensive animal protein in human food can be substituted with soy protein without loss in nutritive value of the diet.

Because of several constraints in the use of dry, mature soybeans, sophisticated protein technology is needed to produce flours, concentrates and isolates, which are further modified into a large number of different products or into meatlike extenders and analogs (7,8,9). Some other workers are beginning to visualize that immature, mature, and germinated soybeans as a vegetable could be an even more universal and versatile protein resource. This has been reflected in the appearance of a number of soybean cookbooks (2,10). Popularization of a vegetarian lifestyle, coupled with studies on the health implications of vegetarian diets, has focused greater attention on soybeans (11). Traditional soy foods based on centuries-old processes developed in the Far East have been described (12,13). In the past 3 years, a soy foods cottage industry has sprung up in the United States to produce some of these foods.

Several reports tell of rapid growth in markets for some traditional oriental foods that have been popularized and Americanized (14,15,16). Soycrafting (17), which emphasizes simple processing technology and household applications, has been influential in this growth.

The purpose of this paper it to illustrate the present availability, acceptability and consumption patterns of soy proteins, to describe the comparative food value of immature, mature and germinated soybeans and to evaluate changes in the antinutritional factors in soybeans with respect to maturation and germination.

Availability and Consumption Patterns of Soy Proteins

Mature Soybeans. The amount of soy protein actually consumed in the United States is less than 3 gram protein per capita per day (1). By contrast, in countries where soybeans are processed at the household level or in small family operations, daily consumption is about 30 gram protein (18). In Japan, where soy foods have a long tradition of use, daily per capita intake in 1975 was 10 gram or 13% of the total protein intake of 79 gram (19).

Although only a small amount of the world's soybean production now is used directly for food, the United States supply of soybeans in crop year 1977 of 1.819 billion bushels (19.8 million metric tons of protein), if used directly for human food, would supply 943 million people with the recommended daily allowance of 56 gram of protein per day for a 70-kg man (Table I). In contrast, the world's production of 2.78 billion bushels (29.4 million metric tons of protein) would provide 1.4 billion people with the recommended allowance of 56 gram protein per day. Several projections indicate that United States production of 3 billion bushels may be needed to satisfy demand for soybeans in 1980. According to Wolf (1), in the United States about 2.5 gram of soy protein/capita/day is consumed directly compared to an available supply of 249 gram soy protein per day. How the U.S. supply of soybeans is disposed of in animal feed, export, and other outlets has been summarized (1).

Table I

Availability and consumption for soy protein in humans

	Soybean Production[a]		Population Equivalent of Protein Supply[b]	Consumption of Soy Protein
	Bushels, Billion	Protein Million Metric Tons	Billion	g/Capita/Day
USA	1.819	19.8	0.943	---
World	2.780	29.4	1.4	---
USA				2.5[a]
Japan				10[c]
Southeast Asia				30[d]

[a]Wolf (1).

[b]Number of people that could be provided with the recommended daily allowance of 56 g protein for a 70-kg man National Research Council (20) from the 1977 soybean crop.

[c]Watanabe (19).

[d]Protein Advisory Group of United Nations (18).

Food Value of Soybeans with Respect to Maturity

Immature Soybeans. Long cooking times are required to improve the texture and flavor of field-dried, mature soybeans, whereas immature beans cook to a tender texture much more quickly. With respect to organoleptic quality, green-mature soybeans may have an advantage because of their softer texture (palatability) and because of the bitter flavor intensity value of only 0.54 as compared to 1.6 at maturity (21). Because of their green color, the only readily apparent potential of immature soybeans to enhance human nutrition is as a vegetable. Furthermore, harvesting difficulties arise because of deteriorative reactions that occur when the tender green beans are bruised or damaged. In the Orient, immature soybeans, known as Endamame, are sometimes used as a substitute for peas and beans; however, extent of usage is not known.

Germinated Soybeans. On the other hand, germinated soybeans may have greater value as a human food. For example, Pomeranz et al. (22) reported that germinated soy flour outperformed a popular dehulled defatted soy flour in bread-making and that the bread made with 10% germinated soy flour was more acceptable by a taste panel. Finney (23) showed that replacement of 7% wheat flour with wet, mashed, freshly germinated soybeans produced breads with no objectionable taste or odor. Germination on the other hand, adversely affected the baking properties of peas and lentils but not faba beans (24,25). However, germination has the potential of generating objectionable flavors since breads containing germinated pea flours had highly unacceptable bitter taste, beany odors and flavors. Blanching eliminated the objectionable flavors but created adverse baking properties (24). Billions of pounds of soybean-cereal blended foods have been shipped to Africa, Latin America, Near East, and Asia (26). Cereal-legume-germinated soy flour combinations could conceivably be used to great advantage in combating malnutrition (27,28).

Since only a few formal taste panel studies have been made, the organoleptic qualities of germinated soybeans will not be discussed. However, germinated soybeans (sprouts), a traditional oriental food, are receiving greater universal attention.

Mature Soybeans. Nutritional value of soy protein has been investigated for several decades. Most of these studies have been concerned with flours, concentrates and isolates derived from dry mature soybeans (4,7,8). Only a few nutritional reports deal with immature and germinated soybeans. The primary purpose of the following sections is to evaluate nutritional value in relation to three edible maturity stages: immature, mature, and germinated soybeans.

Virtually all soybean varieties grown in the United States are derived from six basic genetic lines. Practically all are yellow-seeded varieties, which in turn can be classified as garden-type (vegetable) and field-type soybeans. At present, little differentiation can be made between garden- and field-type soybeans on the basis of chemical composition. Cookability and palatability (flavor and texture) differences may exist (2,29). Analyzed nutrient content of a mature green-seeded, vegetable-type soybean (Verde variety) is consistent with values for yellow-seeded mature varieties (30,31,32).

Compositional Changes During Maturation and Germination. Analyses of fresh weight, percent dry matter, and color of seeds and pods in relation to days after flowering were evaluated for indices of degree of maturity for two soybean varieties (21). One set of results is summarized in Table II. Maximum fresh weight for Hawkeye soybeans was reached at 44 days after flowering, when the seeds are still green and pods begin to yellow. Bates et al. (33) refer to seeds harvested at this stage of maturity as green-mature soybeans.

Oil and Protein. Lipids undergo virtually complete transformation in composition during maturation (34). As shown in Table III, 90% of the total oil found in mature soybeans was synthesized at the maximum fresh weight stage of maturity. There are no further significant changes in saturate:polyunsaturate fatty acid ratios and nutritive value of the oil in green-mature and yellow mature soybeans. At the maximum fresh weight stage of maturity, 61% of protein found in mature soybeans was synthesized (Table IV). On a dry weight basis, percent crude protein increased slightly in going from green-mature to dry, yellow-mature soybeans (35). From these data, it appears that soybeans can be consumed in the green-mature or dry mature form without significant differences in protein and oil quality.

As shown in Fig. 1, very low amounts of extractable protein are found at the earliest stages of maturity. Protein accumulation is very rapid during development and reaches maximum values at about 36 days after flowering, corresponding to maximum fresh weight/seed. High-molecular-weight proteins, which are the predominant storage proteins in mature soybeans, also reach maximum values 36 days after flowering (36,37). The relative amounts of nonprotein nitrogen to protein nitrogen are low at all stages of maturity (35,37).

The major reserve proteins degrade slowly during germination so that very little change occurs in the amino acid composition of soybean sprouts during the first 5 days of germination. For use as a vegetable, soybeans are allowed to germinate for only 4-6 days. That both germination and maturation do not improve the protein quality or eliminate the antinutritional factors in soybeans will be discussed in the following section.

Mineral and Vitamins. Mineral content, expressed on a dry-weight basis, was similar for fresh immature and field-dried mature soybeans. Possible exceptions include higher boron and phosphorus and lower zinc values for meals prepared from mature

Table II

Fresh weight, dry matter and color characteristics of maturing soybeans (1969 crop year)[a]

Days After Flowering	Average Fresh Wt., mg/seed	Dry Matter, %	Color of Beans	
			Pods	Seeds
22	30	16.0	Green	Green
24	59	22.2	Green	Green
27	131	23.0	Green	Green
29	220	23.2	Green	Green
29	295	27.6	Green	Green
31	313	28.0	Green	Green
35	384	30.0	Green	Green
40	498	32.4	Green	Green
44	568[b]	39.0	Yellow	Light-green
49	523	42.7	Yellow	Yellow-green
52	440	49.6	Brown	Yellow
55	331	71.2	Brown	Buff-brown
59[c]	253	84.1	Brown	Buff-brown
64[c]	209	91.6	Brown	Buff-brown

[a]Rackis et al. (21), Hawkeye soybeans.

[b]Maximum fresh weight occurs during the yellowing stage.

[c]Mature beans field-dried before harvest.

Table III

Oil and fatty acid composition of Harosoy 63 soybeans during maturation[a]

Days After	Weight, mg/Seed				Oil Composition, %				
Flowering	Fresh	Dry	Oil,[b] %	mg/Seed	Palmitic	Stearic	Oleic	Linoleic	Linolenic
24	5.5	0.9	3.5	0.03	19.0	6.2	7.5	35.0	30.0
32	146.0	30.8	15.5	4.8	14.5	3.2	24.0	46.2	12.1
39	214.1	64.8	19.5	12.6	10.1	2.8	36.3	51.0	9.2
46	324.0	100.5	20.1	20.7	10.6	3.2	25.8	52.7	7.7
63	452.0	179.0	22.3	39.9	10.4	3.1	26.1	54.1	6.3
72 (maturity)	311.8	186.6	23.6	44.0	10.4	2.9	26.4	54.1	5.8

[a]Rubel *et al*. (35).

[b]Dry basis.

Table IV

Nitrogen composition of Acme soubeans during maturation[a]

Days After Flowering	Weight, mg/Seed		Protein[b]	
	Fresh	Dry	%	mg/Seed
25	22.2	3.8	32.2	1.2
32	124.2	27.3	31.8	8.7
41	266.3	75.7	34.1	25.6
50	370.8	122.2	34.8	42.5
60	359.4	165.6	36.0	59.6
74 (maturity)	238.5	200.6	35.0	70.2

[a]Rubel *et al*. (35).

[b]Dry basis.

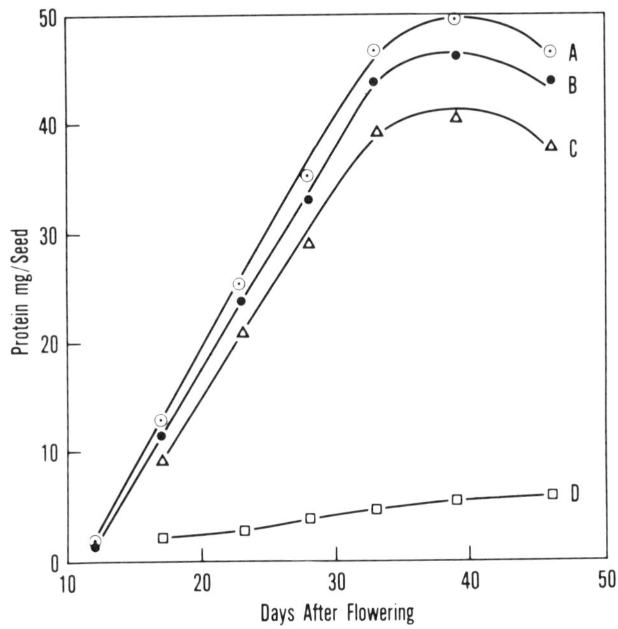

*Figure 1. Changes in protein during maturation. Key: A, total protein; B, extractable protein; C, precipitable protein (0.8*N *trichloroacetic acid); and D, low molecular weight compounds (*36, 37*).*

soybeans compared with meals prepared from immature soybeans (32). Only limited data are available concerning the effect of maturation and germination on the vitamin and mineral content. Compiled data reveal a wide range of values (2).

Changes in ascorbic acid and β-carotene during maturation, storage and subsequent germination are illustrated in Fig. 2. Ascorbic acid and β-carotene in the green-mature stage (maximum fresh weight/seed) are very high. Green-mature, dry mature and sprouted soybeans (5-day germination) contain 30, 2, and 11 mg/100 gram (fresh weight) ascorbic acid, and 0.35, 0.12, and 0.2 mg/100 gram β-carotene, respectively (38). However, appreciable amounts of ascorbic acid and β-carotene are destroyed during the cooking process needed to maximize both protein quality (33) and organoleptic acceptability (2).

Carbohydrates. The amount and nature of the carbohydrates undergo extensive transformation during maturation and germination. As a result, susceptibility to flatulence would be very different. Flatulence and intolerance to other foods have been reviewed elsewhere (39).

Antinutritional Factors and Protein Quality of Soybeans with Respect to Maturity

In the raw state, mature soybeans and many other plant foodstuffs contain protease inhibitors that diminish the proteolytic activities of trypsin and chymotrypsin in the intestinal tract, cause pancreatic hypertrophy and suppress growth. Trypsin inhibitors (TI) account for about 40% of the pancreatic hypertrophic effect and growth-inhibitory capacity of raw soy proteins. The resistance of the raw undenatured protein to tryptic digestion accounts for the remaining 60%. The practical significance of residual TI activity in heat-processed soy protein products and the biochemical effects of other protease inhibitors have been reviewed (40).

Trypsin Inhibitors. TI activity in 108 varieties and strains of soybeans ranged from 66 to 233 trypsin units/mg protein (41). Mean TI values of 57 specific activity units per mg protein were reported for 16 field-type soybeans grown in India, whereas in 8 vegetable-type soybeans, the mean TI value was 41 units/mg protein (42).

As shown in Table V, TI activity generally increased during maturation. The Dare (field-type) variety had the greatest increase in activity. Dehulled immature beans held 2.5 minutes in boiling water had 97-98% of the extractable TI activity destroyed regardless of variety.

Germination for 3 days, after a 24-hour soaking period, did not cause TI activity to change appreciably for the four soybean varieties studied (43). On the other hand, Bates et al. (33) reported that TI activity in water extracts of soybeans germinated 4 days decreased about 70% (Table VI). Autoclaving for

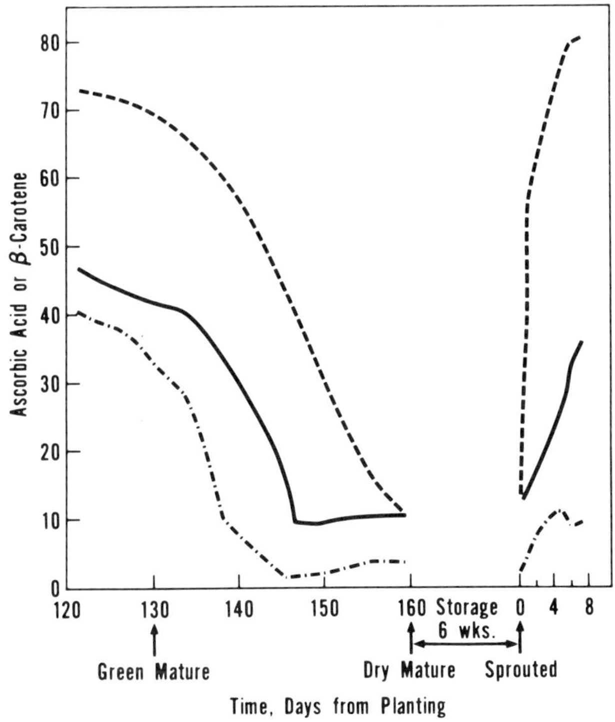

Figure 2. Ascorbic acid, β-carotene, and moisture content of mature and germinated soybeans Bates and Matthews (38). Key: – · – ·, ascorbic acid (mg/100 g fresh or soaked weight); ——, β-carotene (mg/100 g × 100 fresh or soaked weight); and – – –, moisture content.

Table V

Trypsin inhibitor activity of four soybean varieties during maturation and germination[a]

Time Period		Trypsin Inhibitor Activity, mg/g, Dry-Basis[c,d]			
Maturation,[b] Days	Germination Days at 22° C	Soylima[e]	Verde[e]	Kanrich[e]	Dare[f]
1	-	14.0	19.1	21.7	14.4
11	-	16.5	23.0	22.2	20.5
21	-	18.5	23.5	23.6	25.5
28	-	---	---	---	27.4
32	-	17.0	---	---	---
Mature beans	0	17.2	---	22.8	26.9
Soaking	1	16.5	---	22.6	26.9
Germinating	2	16.8	---	21.4	26.4
Germinating	3	16.2	---	20.5	25.2
Germinating	4	16.9	---	19.8	24.0

[a] Collins and Sanders (43).

[b] Harvesting begun early in development with seeds having high moisture content and was discontinued when beans began to dry. Varieties were not at comparable stages of maturity.

[c] Values represent activity extracted under conditions employed and may not represent total activity of the intact seed.

[d] Trypsin inhibitor activity on extracts of dehulled immature beans decreased 97-98% after 2.5 min in boiling water.

[e] Vegetable-type.

[f] Field-type.

Table VI

Effect of maturity on trypsin inhibitor activity and heat inactivation in soybeans[a]

Stage of Maturity	Trypsin Inhibitor Activity, Reaction Rate[b,c] Raw	Heated[d]
Green-mature	49.0	1.5
Mature	52.2	0.6
Sprouts[e]	17.8	1.7

[a]Bates *et al*. (*33*).

[b]Change in absorbance at 257 nm/min/g protein compared to control with no trypsin inhibitor, Bragg variety.

[c]Activity in 0.01N NaOH extracts.

[d]Autoclaved 121° C for 15 min.

[e]Soaked 6 h and germinated at 27° C for 4 days.

15 minutes at 121°C destroyed nearly all of the TI's in immature, mature and germinated soybeans.

Special mention should be made concerning the TI activity values of immature and germinated soybeans mentioned in this report (Tables V and VI). TI activity extracted under these conditions may not represent total activity of the intact bean (44,45). Also, much of the TI data obtained by several workers cannot be compared directly because methodology differed widely and TI activity units are not interchangeable. TI activity in extracts of immature and germinated soybeans is readily destroyed by autoclaving and other forms of moist heat treatment. Undoubtedly, most of the total TI in intact immature and germinated soybeans would be readily destroyed by heat treatment, since moisture content is very high. This conclusion is derived from the fact that in whole mature soybeans containing about 10% initial moisture, only 10% of the total trypsin inhibitor activity was destroyed during steaming at 100°C for 20 minutes. When the mature soybeans were first soaked or tempered to at least 25% moisture content, over 97% TI activity was destroyed during heat treatment for 20 minutes (46). Other studies on the importance of particle size on rate of destruction of TI activity with moist heat have been reported by Albrecht et al. (47) and Baker and Mustakas (48).

None of the animal feeding studies with immature and germinated soybeans included an evaluation of pancreatic enlargement (40). Maximum growth occurs in rats fed soy diets in which only 79% of the TI activity was eliminated by live steam treatment, and pancreatic hypertrophy did not occur with only 54% destruction of TI (49). Biochemical threshold level effects in long-term rat feeding studies confirm these findings (50). Therefore, it is highly unlikely that pancreatic hypertrophy would occur in animals fed immature and germinated soybeans, since the TI's are readily destroyed by moist heat treatment.

Several different forms of protease inhibitors have been isolated from soybeans, and several genetic variants of the two major soy protease inhibitors (Kunitz trypsin inhibitor and Bowman-Birk inhibitor) have been identified. A new form of TI immunochemically identical to the Kunitz TI appeared during germination of six populations of soybeans. These studies have been reviewed elsewhere (40). In spite of differences in level of protease activity and the presence of protease inhibitor variants, heat treatment was effective in obtaining maximum PER and in eliminating pancreatic hypertrophy in soybean meals prepared from several varieties and strains of mature soybeans (41).

Protein Quality. The protein quality of soybeans at different stages of maturity is shown in Table VII. Sulfur-amino acid content was independent of maturity. In the raw form, protein quality was very low for immature, mature, and germinated soybeans. After heat treatment, PER values were increased to

Table VII

Protein quality of immature, mature and germinated soybeans (Bragg variety)[a]

Maturity	Composition, Dry-Basis Protein, %	Composition, Dry-Basis Lipid, %	Methionine, g/16 g N	Cystine, g/16 g N	PER[b]
Green mature, raw	36.7	20.5	1.22	0.59	0.77
Dry mature, raw	39.4	21.0	1.18	0.61	0.75
Sprouts, raw	41.7	22.3	1.19	0.63	0.64
Green mature, heated	---	---	---	---	2.05
Dry mature, heated	---	---	---	---	2.11
Sprouts, heated	---	---	---	---	2.02
Casein	---	---	2.46	0.21	2.50

[a]Bates *et al.* (33).

[b]Protein efficiency ratio corrected on a basis of PER = 2.50 for casein, no significant difference in PER in respect to stage of maturity when compared in either the raw or heated form.

values 80% of that for casein. The practical significance of the heat inactivation data given in Tables V-VII is that no stable forms of TI exist in soybeans, regardless of maturity, and that other antinutritional substances that may be present can be readily eliminated by heat treatment or may exist at levels that have no effect on nutritional quality as evidenced in short-term feeding tests (40,50).

Protein Digestibility. The reported beneficial effects of germination on soy protein quality are conflicting and often contradictory. Few conclusions can be drawn from the large amount of reported data for other food legumes. Some investigators found an increase in protein quality and other nutrients, others indicated no change or a decrease in nutritive value, and still others found the data were inconclusive. Some of the more definitive references on protein quality and digestibility, mineral bioavailability, phytic acid, and trypsin inhibitors include: (51-63). Finney (23) and Hsu et al. (24) cite many references that indicate germination of wheat may increase mineral availability, lysine and several vitamins, improve bread quality and significantly increase relative nutritive value of many cereals, whereas Lorenz (64) concludes little change in nutrients occurs.

In the initial stages of germination there is a marked increase of proteolytic enzyme activity in the cotyledons of soybeans and other food legumes, followed by a decline. However, there is no evidence that the altered amino acid pattern of ungerminated and 5-day germinated seeds is enough to affect protein efficiency ratios (PER) (Table VII). Concomitant with proteolysis are increases in nonprotein nitrogen, free amino acid nitrogen, uric acid and ureide nitrogen in cotyledons and embryo during germination (65,66).

Since little data is available for soybeans, one must turn to studies on other food legumes to gain insight on the effects of germination on protein quality. Digestibility values for two food legumes are given in Table VIII. Germination greatly increased digestibility of raw red kidney beans, and cooking further increased protein digestibility. Digestibility values of cooked germinated kidney beans were significantly higher than values for cooked ungerminated kidney beans (84.4 vs. 69.3%, respectively). The beneficial effect of germination on protein digestibility is attributed to a 50% decrease in trypsin inhibitor activity and an increase in digestibility of the reserve storage globulins. On the other hand, germination appears to lower digestibility values of cooked common beans.

Soybeans present an intermediate pattern with respect to the effect of germination on protein value of the beans, in that PER values of ungerminated and germinated soybeans are not significantly different (Table VII). Apparently, a slower proteolytic degradation of the major reserve proteins in germinating soybeans had little effect on protein digestibility.

Table VIII

Digestibility values for germinated legumes and globulin proteins

	Digestibility Values, %		
		Cooked	
Dietary Protein	Raw Red Kidney Beans[a]	Red Kidney Beans	Common Beans[b]
Casein[c]			
Ungerminated beans	29.5	69.3[d]	67[e]
Germinated beans	66.4[f]	84.4	60[g]
Isolated globulins			
Ungerminated beans	62.5	71.6	--
Germinated beans	73.4[f]	82.6	--

[a] *Phaseolus vulgaris* El-Hag *et al.* (63).

[b] *Phaseolus vulgaris*, S-19N variety Elias *et al.* (58).

[c] Digestibility value = 91.4.

[d] All fractions cooked 15 min at 121° C.

[e] All beans cooked 10 min at 121° C.

[f] 10-Day germination.

[g] 9-Day germination.

In contrast to most studies, Elias et al. (58) reported that PER values in cooked common beans (Phaseolus vulgaris, variety S-19n) decreased from 0.99 in intact beans to 0.59 and 0.26 after 6- and 9-day germination, respectively. Apparent digestibility values for sprouted common beans decreased (Table IX) after 9 days of germination, and trypsin inhibitor activity also decreased appreciably. The lower nutritive value of germinated common beans correlated with a decrease in total sulfur amino acids (58). After 3 days of germination, PER values decreased in both raw and cooked forms of green gram, cowpeas and chick peas.

The results of studies reviewed here, as well as those reported by other workers, are conflicting. Depending upon the food legume, germination may either improve or decrease nutritive value or have little or no beneficial effect. Aside from differences in varieties and conditions used in germinating seeds, proteolytic activity during germination is probably the most important factor that may affect nutritive value by modifying protein digestibility and amino acid profile of the limiting essential amino acids, methionine, and cystine. Sprouting of cereals causes changes in nutrients which may not represent true increases but rather reflect loss of dry matter primarily as carbohydrate. Animal studies fail to show a beneficial effect of sprouting on the nutritive value of cereals (64).

Phytic Acid. Recent reviews (67,68,69) summarized the literature covering the relationship between phytic acid and mineral bioavailability in soy protein products. The formation of phytate-protein-mineral complexes (particularly zinc chelates in flours, concentrates, and isolates prepared from mature soybeans) may be responsible for reduced mineral availability. However, the iron in ^{59}Fe-labeled mature soybeans is more available to iron-deficient rats than the iron in green-immature soybeans, even though mature soybeans contain three times more phytic acid (70). The factor(s) responsible for this difference in bioavailability has not been identified.

Numerous animal studies suggest that bioavailability of minerals from diets containing high phytate levels varies considerably. Complex interactions between the minerals and phytic acid contained in foodstuffs, intestinal and food phytase activities, food processing conditions, protein digestibility, and physiological conditions in the intestinal tract are factors that must be considered in predicting mineral bioavailability. Dietary fiber along with phytic acid has been cited as a possible cause for low mineral bioavailability of cereals and oilseeds, and yet soybean hulls have no significant effect upon the availability of the soy flour zinc or of the calcium added to rats fed soy diets (71).

Seed germination decreases phytate and increases phytase activity (Table X). A 22% decrease in phytic acid occurs during 5 days of soybean germination (62). Phytase activity in soybeans increased 227% compared to an increase of 907-3756% in peas.

Table IX

Effect of germination on nutritive value of selected legumes[a]

Days of Germination	Basal Diet	Common Bean[b]	Green gram[c]	Cowpea[d]	Chick Pea[e]
		Protein Efficiency Ratios			
-	Casein	---	2.32	2.34	2.45
0		0.99	(1.16)[f] 1.35	(0.78) 1.55	(1.99) 2.47
3		0.86	(1.02) 1.10	(0.80) 1.20	(1.67) 2.28
6		0.59	---	---	---
9		0.26	---	---	---

[a]Common bean Elias *et al.* (58); other beans Jaya *et al.* (72).

[b]*Phaseolus vulgaris*.

[c]*Phaseolus aureus*.

[d]*Vigna sinensis*.

[e]*Cicer arietinum*.

[f]Values in parentheses represent values for raw beans.

Table X

Changes in phytate content and phytase activity during germination of soybeans and peas[a,b]

	Phytic Acid		Phytase	
			Activity[c]	
Source	mg P/g Sample[b]	% Change	mg P Released/g	% Change
Soybean seed	2.48	--	0.30	---
Soybean sprouts	1.94	22	0.68	227
Dwarf gray peas	1.13	--	0.04	---
Pea sprouts	0.59	48	0.39	907
Early Alaska peas	1.86	--	0.02	---
Pea sprouts	1.20	35	0.65	3,756

[a]Chen and Pan (62).

[b]5-Day germination.

[c]At pH 5.2, 48 h, 37°C.

Plant phytases have all of the characteristics of a nonspecific acid phosphomonoesterase with broad substrate specificity (73).

Hypercholesterolemia. Diet has long been considered to play an important role in hypercholesterolemia and atherosclerosis, but most of the emphasis has been on dietary fat and cholesterol. Several clinical and experimental studies demonstrate that replacement in the diet of animal protein with soy protein had a striking hypocholesterolemic effect (74,75).

The hypocholesterolemic activity of many food legumes can be enhanced when germinated (76,77). Presumably an increase in certain isoflavones during germination may account for the lowering of serum cholesterol levels (78). Plant sterols which also have been implicated as hypocholesterolemic dietary factors increase during germination and large changes in their chemical composition occur (79).

Estrogens. Concern over dietary estrogens has focused largely on consumption of residual diethylstilbestrol (DES) in animal tissue fed the compound as a growth promoter. Human exposure to naturally-occurring phytoestrogens, and adventitious estrogens derived from microbial organisms, insecticides, and drugs can be substantially higher than exposure to DES residues (80). The significance of long-term dietary exposure to estrogen in amounts smaller than physiological or pharmacological doses is highly speculative.

To avoid hazards resulting from the ingestion of naturally-occurring toxicants in various foodstuffs, Coon (81) advanced the concept of "Safety in Numbers." That is, the wider the variety of food intake, the less chance any one constituent can reach a hazardous level in the diet. Another concept "Safety through Technology" can be practiced, by which food processing is used to destroy or eliminate the toxicant.

Many plants possess estrogenic activity. Literature on phytoestrogens can be classified into three general categories: isoflavones, coumestans and resorcylic acid lactones (80). Estrogenic activity has been detected in many foodstuffs (Table XI). However, the presence in plants of steroidal estrogens such as estradiol, estriol, and estrone could not be confirmed by mass fragmentographic analysis (82). Estrogenic activity is based mostly on evidence of uterine enlargement or of cornification of vaginal epithelium in experimental animals. Coumestrol content of plant foodstuffs is given in Table XII.

The 8- to 197-fold increase in coumestrol concentration with germination time may reflect inherent varietal differences and environmental conditions during the growing season, as well as microbial effects. Whether these factors or the methodology employed for germination may explain the wide differences in reported values for coumestrol content in soybean sprouts by Knuckles et al. (83) and Lookhart et al. (84) remains to be evaluated. Dehulling ungerminated and germinated soybeans can

Table XI

Estrogenic activity of common plant foodstuffs

Foodstuff	Amount[a]
Carrots, fresh	+
Cabbage	0.024 μg E_2/g[b]
Peas	0.004-0.006 μg E_2/g
Hops	1-300 μg E_2/g
Wheat bran	+
Wheat germ	+
Rice bran	+
Rice polish	+
Soybean meal	+
Vegetable oils	+
Pomegranate seeds	4-17 mg estrone/kg
Milk	+

[a]Compiled by Verdeal and Ryan (80) from various sources.

[b]E_2 = 17-β-estradiol.

Table XII

Coumestrol content of plant products

Product	Coumestrol, ppm, Dry Basis A[a]	Coumestrol, ppm, Dry Basis B[b]
Alfalfa sprouts	5.0	-
Soybean sprouts	71.1	0.23-3.94[c,d]
Soybeans	1.2	0.02-0.09[c,d]
Soybean meal (defatted)	0.4	-
Soy protein concentrate	0.2	-
Soy protein isolate	0.6	-
Green beans (frozen)	1.0	-
Snow peas (frozen)	0.6	-
Green peas (frozen)	0.4	-
Brussell sprouts	0.4	-
Red beans	0.4	-
Split peas	0.3	-
Spinach leaf (frozen)	0.1	-

[a]Knuckles _et al_. (_83_).

[b]Lookhart _et al_. (_84_).

[c]Values for three varieties, corrected for the 36% unextractable coumestrol Lookhart _et al_. (_85_).

[d]Some varieties contain much higher levels (Lookhart personal communication).

reduce total coumestrol content 37-63% (85). Coumestrol content of commercial flours from germinated and ungerminated soybeans has been reported (86).

As indicated in Table XIII, human exposure to soybean isoflavones, measured in DES equivalents, is considerably more likely than exposure to coumestrol in soybeans. Several fold increases in isoflavone content during germination is a definite possibility. The importance of soybean estrogens, particularly in germinated soybeans, in the etiology of deleterious reactions in humans needs more study. Soybean isoflavones and phenolic acids probably account for the antioxidant activity of soybeans, defatted soy flour, protein concentrates, and isolates (89).

Ohta et al. (90) isolated two new isoflavones from soybeans, one of which was identified as 6"-O-acetyl daidzin; glycitein and glycitein-7-O-glucoside, which were isolated by Naim et al. (87), were not detected by these workers.

Conclusions

Although many food legumes in the immature, mature, and germinated stages can vary widely in nutritive value, no readily apparent information can account for the fact that nutritive value, and protein quality in particular, can either increase, decrease, or remain the same. On the other hand, protein quality of immature, mature, and germinated soybeans is not significantly different when properly processed with moist heat treatment. To eliminate the growth-inhibiting and pancreatic hypertrophic effects of raw soybeans, regardless of the state of maturity, cooking is required to inactivate protease inhibitors and convert raw proteins into more readily digestible forms.

There may be some nutritional benefits in consuming immature and germinated soybeans compared to products prepared from mature soybeans. For example, ascorbic acid and β-carotene are present in rather high amounts in immature and germinated soybeans, whereas mature soybeans are practically devoid of these vitamins. Content of some of the other vitamins and minerals may differ appreciably, but the general composition of soybeans at the three stages of maturity is essentially the same.

Better cooking and taste characteristics are the main advantages of direct consumption of green-immature and germinated soybeans as a vegetable. Mature soybeans require long cooking times to make them tender. Organoleptic qualities of immature beans are reported to be much better than those of mature soybeans or of flours, concentrates, and isolates prepared from them. Flours from germinated soy beans may have better bread-making properties than flours made from mature soybeans.

The nutritional properties of acceptable foods from soybeans at various stages of maturity should encourage the development of soybeans as an important and versatile international protein resource. Present world production of soybeans, if used directly for human consumption, can supply 1.4 billion people with the

Table XIII

Human exposure to exogenous estrogens

Estrogen Source	Estimate of Possible Daily Dose (μg/DES[a] Equivalents)
Morning-after pill	50,000[b]
Birth control pill	2,500[b]
100 g Beef liver with 0.5 ppb DES	0.05[b]
100 g Wheat with 2 ppm zearalenone	0.2[b]
100 g Beans (*Phaseolus vulgaris*) with 2-10 ppb estradiol	0.03-0.15[b]
20 g (Dry basis) soybean sprouts with 70 ppm coumestrol	0.5[b]
Soybeans with 1.2 ppm coumestrol	0.0085[b]
100 g Soybean meal with 164.4 mg[b,c] genistin	1.64[d]
100 g soybean meal with 1.4 mg genistein[c]	0.014[d]
100 g Soybean meal with 58.1 mg[c] daidzin	0.04[d]
100 g Soybean meal with 0.6 mg daidzein[c]	0.0045[d]
100 g Soybean meal with 0.11 mg glycitein[c]	e
100 g Soybean meal with 33.8 mg glycitein-7-β-O-glucoside[c]	e

[a]DES = diethylstilbestrol.

[b]Compiled by Verdeal and Ryan (80).

[c]Naim *et al*. (87).

[d]Calculated according to relative potency in mice Bickoff *et al*. (88).

[e]Estrogenic potency unknown.

recommended daily allowance of 56 g protein for a 70-kg man. In addition to the nutritional data reviewed, it is necessary to consider the difficulties of harvesting and storing immature soybeans. Special harvesting and shelling equipment is needed to prevent damage (91,92). Costs associated with germination and spoilage problems during storage of immature soybeans also are important considerations. More research is needed to establish optimal conditions of germination in order to provide products with acceptable functional and organoleptic properties. Germination conditions can adversely affect germination activity and vigor, flavor, microbial activity, and wholesomeness of the final product (23,24).

Literature Cited

1. WOLF, W. J. in "The World Soybean Research Conference - II, Proceedings" (F. T. Corbin, ed.), pp. 761-780). Westview Press, Inc., Boulder, Colorado. (1980)
2. RACKIS, J. J. in "Postharvest Biology and Biotechnology." (H. O. Hultin and M. Milner, eds.), pp. 34-76. Food and Nutrition Press, Inc., Westport, Connecticut. (1978)
3. WOLF, W. J. in "Proc. Int. Soya Protein Food Conf.," Singapore, January 25-27, pp. 59-65. American Soybean Association. (1978)
4. WILCKE, H. L.; HOPKINS, D. T. and WAGGLE, D. H. (eds.). "Soy Protein and Human Nutrition," Academic Press, New York. (1979)
5. SARETT, H. P. in "World Soybean Research." Proceedings of World Soybean Research Conference, (L. D. Hill, ed.), pp. 840-849. The Interstate Printers and Publishers, Inc., Danville, Illinois. (1976)
6. YOUNG, V. R.; SCRIMSHAW, N. S.; TORUN, B.; VITERI, F. J. Am. Oil Chem. Soc. 1979, 56, 110-120.
7. SMITH, A. K. and CIRCLE, S. J. (eds.). "Soybeans: Chemistry and Technology, Vol. II, Proteins." Avi Publishing Co., Westport, Connecticut. (1979)
8. AMERICAN SOYBEAN ASSOCIATION. "Proceedings of the World Soy Protein Conference," Munich, Germany, November 11-14, 1973. J. Am. Oil Chem. Soc. (1974) 51, 49A-207A.
9. AMERICAN OIL CHEMISTS' SOCIETY. "Proceedings World Conference on Vegetable Food Proteins," Amsterdam, The Netherlands, October 29-November 3, 1978. J. Am. Oil Chem. Soc. (1979) 56, 99-483.
10. RACKIS, J. J. in "Tropical Foods: Chemistry and Nutrition," Vol. 2. (G. E. Inglett and G. Charalambous, eds.), pp. 485-510. Academic Press, New York. (1979)
11. PHILLIPS, R. L. Cancer Lett. 1979, 35, 3513-3522.
12. HESSELTINE, C. W.; WANG, H. L. in "Soybeans: Chemistry and Technology, Vol. I, Proteins." (A. K. Smith and S. J. Circle, eds.), pp. 389-419. Avi Publishing Co., Westport, Connecticut. (1972)

13. AMERICAN SOYBEAN ASSOCIATION. "Proceedings of the International Soya Protein Food Conference," Singapore, January 25-27, 1978. American Soybean Association, St. Louis, Missouri. (1978)
14. SHURTLEFF, W. and AOYAGI, A. (eds.). "The Book of Tempeh." Harper and Row Publishers, Inc., New York. (1979)
15. SHURTLEFF, W. and AOYAGI, A. (eds.). "Tofu and soymilk production. The Book of Tofu," Vol. II, New-Age Foods Study Center, Lafayette, California. (1979)
16. BAUER, C.; ANDERSON, J. (eds). "The Tofu Cookbook," Rodale Press, Emmaus, Pennsylvania. (1979)
17. LEVITON, R. (ed.). Soyfoods-The. Journal of the Soycrafters Association of North America, Ashuelot Valley Printing, Keene, New Hampshire.
18. PROTEIN ADVISORY GROUP OF THE UNITED NATIONS. PAG Statement No. 22, PAG Bulletin, Vol. III, No. 2, United Nations, New York. (1973)
19. WATANABE, T. in "Proc. Int. Soya Protein Conf.," Singapore, January 25-27, American Soybean Association, (1978), pp. 35-38.
20. NATIONAL RESEARCH COUNCIL. "Recommended Dietary Allowances," 8th revised edition, National Academy of Science, Washington, D.C. (1974)
21. RACKIS, J. J.; HONIG, D. H.; SESSA, D. J.; MOSER, H. A. Cereal Chem. 1972, 49, 586-597.
22. POMERANZ, Y.; SHOGREN, M. D.; FINNEY, K. F. J. Food Sci. 1977, 42, 824-827, 842.
23. FINNEY, P. L. in "Nutritional Improvement of Food Proteins." (M. Friedman, ed.), Advances in Experimental Medicine and Biology Series, Vol. 105, Plenum Press, New York. (1979)
24. HSU, D.; LEUNG, H. K.; FINNEY, P. L.; MORAD, M. M. J. Food Sci. 1980, 45, 87-92.
25. FINNEY, P. L.; MORAD, M. M.; HUBBARD, J. D. Cereal Chem. in press.
26. AGENCY FOR INTERNATIONAL DEVELOPMENT. Food for Peace, PL-480, Title II, Commodities Reference Guide, U.S. Department of State, Washington, D.C., November 15. (1976)
27. AGENCY FOR INTERNATIONAL DEVELOPMENT. An inventory of information on the utilization of unprocessed and simply processed soybeans as human food. Contract AID/AG/TAB-225-12-76. U.S. Department of State, Washington, D.C. (1976)
28. BRESSANI, R.; ELIAS, L. G. in "New Protein Foods." (A. M. Altschul, ed.), pp. 230-297. Vol. 1A, Technology, Academic Press, New York. (1974)
29. PERRY, A. K.; PETERS, C. R.; VAN DUYNE, F. O. J. Food Sci. 1976, 41, 1330-1334.
30. CRITTENDEN, H. W.; RAHN, E. M.; WISK, E. L.; WOODMAN, G. W. The Verde Soybean. Del. Agric. Exp. Sta. Circ. No. 9. (1968)

31. SVEC, L. in "Final Report--An Interdisciplinary Program on the Characterization and Evaluation of Edible Green Seeded Soybeans." (P. R. Austin, ed.) University of Delaware Research Foundation, Newark, Delaware. (1970)
32. RASMUSSEN, A. I. J. Am. Diet. Assoc. 1978, 72, 604-608.
33. BATES, R. P.; KNAPP, F. W.; AVAUJO, P. E. J. Food Sci. 1977, 42, 271-272.
34. PRIVETT, O. S.; DOUGHERTY, K. A.; ERDAHL, W. L.; STOLYHEVO, A. J. Am. Oil Chem. Soc. 1973, 50, 516-520.
35. RUBEL, A.; RINNE, R. W.; CANVIN, D. T. Crop Sci. 1972, 12, 739-741.
36. HILL, J. E.; and BREIDENBACH, R. W. Plant Physiol. 1974, 53, 742-746.
37. HILL, J. E.; BREIDENBACH, R. W. Plant Physiol. 1974, 53, 747-751.
38. BATES, R. P.; MATTHEWS, R. F. Proc. Fla. State Hort. Soc. 1975, 88, 266-271.
39. RACKIS, J. J. J. Am. Oil Chem. Soc. in press.
40. RACKIS, J. J.; GUMBMANN, M. R. in "Antinutrients and Natural Toxicants." (R. L. Ory, ed.), Chapter 12. Food and Nutrition Press, Inc., Westport, Connecticut. (1981)
41. KAKADE, M. L.; SIMONS, N. R.; LIENER, I. E.; LAMBERT, J. W. J. Agric. Food Chem. 1972, 20, 87-90.
42. GUPTA, A. K.; DEODHAR, A. D. Indian J. Nutr. Diet. 1975, 12, 81-84.
43. COLLINS, J. L.; SANDERS, G. G. J. Food Sci. 1976, 41, 168-172.
44. RACKIS, J. J.; McGHEE, J. E.; LIENER, I. E.; KAKADE, M. L.; PUSKI, G. Cereal Sci. Today 1974, 19, 513-516.
45. KAKADE, M. L.; RACKIS, J. J.; McGHEE, J. E.; PUSKI, G. Cereal Chem. 1974, 51, 376-382.
46. RACKIS, J. J. J. Am. Oil Chem. Soc. 1974, 51, 161A-174A.
47. ALBRECHT, W. J.; MUSTAKAS, G. C.; McGHEE, J. E. Cereal Chem. 1966, 13, 400-407.
48. BAKER, E. C.; MUSTAKAS, G. C. J. Am. Oil Chem. Soc. 1973, 50, 137-141.
49. RACKIS, J. J.; McGHEE, J. E.; BOOTH, A. N. Cereal Chem. 1975, 52, 85-92.
50. RACKIS, J. J.; McGHEE, J. E.; GUMBMANN, M. R.; BOOTH, A. N. J. Am. Oil Chem. Soc. 1979, 56, 162-168.
51. EVERSON, G. J.; STEENBOCK, H.; CEDERQUIST, D. C.; PARSONS, H. T. J. Nutr. 1944, 27, 225-229.
52. MATTINGLY, J. P.; BIRD, H. R. Poultry Sci. 1945, 24, 344-352.
53. DESIKACHAR, H. S. R.; DE, S. S. Science 1947, 106, 421-422.
54. CHATTAPADHGAY, H. S.; BANNERJEE, S. Indian J. Med. Res. 1953, 41, 185-189.
55. KAKADE, M. L.; EVANS, R. J. J. Food Sci. 1966, 31, 781-783.
56. ABU-SHAKRA, S.; MIRZA, S.; TANNOUS, R. J. Sci. Food Agric. 1970, 21, 91-93.

57. PUSZTAI, A. Planta 1972, 107, 121-129.
58. ELIAS, L. G.; CONDE, A.; MUNZ, A.; BRESSANI, R. in "Nutritional Aspects of Common Beans and Other Legume Seeds as Animal and Human Foods." (W. G. Jaffee, ed.), pp. 139-163. Archivos Latinoamericanos de Nutricion, Caracas, Venezuela (1973)
59. PALMER, R.; McINTOSH, A.; PUSZTAI, A. J. Sci. Food Agric. 1973, 24, 937-944.
60. CHEN, L. H.; WELLS, C. E.; FORDHAM, J. R. J. Food Sci. 1975, 40, 1290-1294.
61. FORDHAM, J. R.; WELLS, C. E.; CHEN, L. H. J. Food Sci. 1975, 40, 552-556.
62. CHEN, H.; PAN, S. H. Nutr. Rep. Int. 1977, 16, 125-131.
63. EL-HAG, N.; HAARD, N. S.; MORSE, R. E. J. Food Sci. 1978, 43, 1874-1875.
64. LORENZ, K. CRC Critical Reviews in Food Science and Nutrition 1980, 13, 353-385.
65. THAPAR, V. K.; PAUL, B.; SINGH, R. Plant Biochem. J. 1974, 1, 11-15.
66. CHEN, L. H.; THACKER, R. J. Food Sci. 1978, 43, 1884-1885.
67. RACKIS, J. J.; ANDERSON, R. L. Food Prod. Dev. 1977, 11(10), 38, 40, 44.
68. ERDMAN, Jr., J. W. J. Am. Oil Chem. Soc. 1979, 56, 736-741.
69. CHERYAN, M. CRC Critical Reviews in Food Science and Nutrition 1980, 13, 297-336.
70. WELCH, R. M.; CAMPEN, D. R. J. Nutr. 1975, 105, 253-256.
71. WEINGARTNER, K. E.; ERDMAN, Jr., J. W.; PARKER, H. M.; FORBES, R. M. Nutr. Rep. Int. 1979, 19, 223-231.
72. JAYA, T. V.; KRISHNAMURTHY, K. S.; VANKATARAMAN, L. V. Nutr. Rep. Int. 1975, 12, 175-183.
73. LOLAS, G. M.; MARKAKIS, P. J. Food Sci. 1977, 42, 1094-1097, 1106.
74. SIRTORI, C. R. et al. Am. J. Clin. Nutr. 1979, 32, 1645-1658.
75. CARROLL, K. K.; HUFF, M. W.; ROBERTS, C. K. in "Soy Protein and Human Nutrition." (H. L. Wilcke, D. T. Hopkins, D. H. Waggle, eds.), pp. 261-280. Academic Press, New York. (1979)
76. JAYA, T. V.; VENKATARAMAN, L. V. Nutr. Rep. Int. 1979, 20, 383-392.
77. JAYA, T. V.; VENKATARAMAN, L. V.; KRISHNAMURTHY, K. S. Nutr. Rep. Int. 1979, 20, 371-381.
78. SHARMA, R. D. Lipids 1979, 14, 535-540.
79. YOSHIDA, H.; KAJIMOTO, G. Agric. Biol. Chem. 1979, 43, 705-712.
80. VERDEAL, K.; RYAN, D. S. J. Food Prot. 1979, 42, 577-583.
81. COON, J. M. in "Toxicants occurring naturally in foods," 2nd Ed. Committee on Food Protection, Food and Nutrition Board. National Research Council, National Academy of Science, pp. 573-594. Washington, D.C. (1973)

82. VAN ROMPUY, L. L. L.; ZEEVAART, J. A. D. Phytochemistry 1979, 18, 863-865.
83. KNUCKLES, B. E.; de FREMERY, D.; KOHLER, G. O. J. Agric. Food Chem. 1976, 24, 1177-1180.
84. LOOKHART, G. L.; FINNEY, P. L.; FINNEY, K. F. Cereal Chem. 1979, 56, 495-496.
85. LOOKHART, G. L.; JONES, B. L.; FINNEY, K. F. Cereal Chem. 1979, 56, 386-388.
86. LOOKHART, G. L.; FINNEY, K. F.; FINNEY, P. L. in "Liquid Chromatography." (G. Charalambous, ed.), pp. 129-138. Academic Press, New York. (1979)
87. NAIM, M.; GESTETNER, B.; ZILKAH, S.; BIRK, Y.; BONDI, A. J. Agric. Food Chem. 1974, 22, 806-810.
88. BICKOFF, E. M.; LIVINGSTON, A. L.; HENDRICKSON, A. P.; BOOTH, A. N. J. Agric. Food Chem. 1962, 10, 410-412.
89. PRATT, D. E.; BIRAC, P. M. J. Food Sci. 1979, 44, 1720-1722.
90. OHTA, N.; KUWATA, G.; AKAHORI, H.; WATANABE, T. Agric. Biol. Chem. 1979, 43, 1415-1419.
91. COLLINS, J. L.; Mc CARTY, I. E. Tenn. Home Sci. Prog. Rep. No. 69, 1969, Knoxville, Tenn., pp. 1-4.
92. COLLINS, J. L.; Mc CARTY, I. E.; SWINGLE, H. D. Tenn. Home Sci. Prog. Rep. No. 78, 1971, Knoxville, Tenn., pp. 1-4.

RECEIVED April 15, 1981.

16

Lipid Oxidation in Fruits and Vegetables

ROLAND TRESSL, DAOUD BAHRI, and KARL-HEINZ ENGEL

Technische Universität Berlin, Seestrasse 13, D-1000 Berlin 65, West Germany

The enzymic formation of aldehydes, ketones, alcohols, and oxoacids (from linoleic and linolenic acids) on disruption of plant tissues is an important biosynthetic pathway by which fruit and vegetable volatiles are formed. Some examples are: (E)-2-hexenal ("leaf aldehyde") and (Z)-3-hexenol ("leaf alcohol") in tea; (E)-2-hexenal in apples; (E,Z)-2,6-nonadienal ("violet leaf aldehyde") and (E)-2-nonenal in cucumber; (Z)-6-nonenal in musk melon; (Z,Z)-3,6-nonadienol in water melon, and 1-octen-3-ol ("mushroom alcohol") in certain edible mushrooms and Fungi. The enzyme system is highly substrate specific to a (Z,Z)-1,4-pentadiene system (like lipoxygenase) splitting the $>C=C<$ double bond at the W - 6 and/or W - 9 position. Therefore linoleic-, linolenic-, and arachidonic acids are natural substrates. It seems to be a common principle in leaves, fruits, vegetables, and basidiomycetes. Some of the volatiles formed are known as important aroma components, pheromonones or wound hormones.

In 1966 our experiments with fruit homogenates showed that linolenic acid is transformed into (Z)-3-hexenal, (E)-2-hexenal and the corresponding alcohols. By means of radio labeling experiments with ripe bananas we could demonstrate that the precursor is converted into (E)-2-hexenal and 12-oxo-(E)-10-dodecenoic acid. Green bananas decomposed (U-^{14}C)-linolenic acid into (E,Z)-2,6-nonadienal and 9-oxononanoic acid (1, 2).

Figure 1 shows C_6-, C_9-, C_8-, C_{10}-, and C_{12}-components which are formed from linolenic acid in certain plants, fruits, vegetables, and mushrooms. There are normally four enzymes involved which seem to be membrane bonded and located in the chloroplasts. In 1970 we tried to isolate and characterize these enzymes and could demonstrate that lipoxygenase (E_1), an

0097-6156/81/0170-0213$05.00/0

alcohol oxidoreductase (E_4), and a (*Z*)-3 / (*E*)-2-isomerase (E_3) are operative in ripe and green bananas and apples. We failed to isolate an active enzyme E_2 converting the 13-LOOH into hexenal (and C_{12}-oxoacid) or 13-LnOOH into (*Z*)-3-hexenal and (*E*)-2-hexenal. This was pretty frustrating because the enzyme E_2 showed a very high activity in homogenates and tissue slices but no activity at all in the isolates. We stopped our experiments and called E_2 = "aldehyde lyase". In 1976 Vick and Zimmerman (3) demonstrated a "hydroperoxide lyase" activity in water melon seedlings and in 1977 - 1978 Galliard and coworkers (4, 5, 6) characterized a "hydroperoxide cleaving enzyme" from fruits of cucumbers and tomatoes. The cleaving enzyme E_2 was located in the chloroplasts and other subcellular membranes and lost its activity during purification. 13- and 9-LOOH were good substrates but not the corresponding hydroxy acids or α-ketols. On the other hand Hatanaka and coworkers (7) investigated an enzyme system in *Thea sinensis* chloroplasts which converted (U-^{14}C)-linolenic acid into (*Z*)-3-hexenal, (*E*)-2-hexenal, and 12-oxo-(*Z*)-9-, and 12-oxo-(*E*)-10-dodecenoic acids. (*Z,Z*)-3,6-nonadienoic acid was also converted into (*Z*)-3-hexenal. Therefore E_2 in *Thea* chloroplasts seems to be no lipoxygenase system. Many plants (and fruits) possess lipoxygenase but no cleaving enzymes E_2. The hydroperoxides are converted into carbonyls, alcohols, and oxoacids by chemical reactions (examples: cooked asparagus, potatoes, malt). The products in these systems are comparable to those of the autoxidation of fatty acids.

Formation of Carbonyls and Oxoacids by Chemical Reactions

Thermal oxidation, autoxidation, and light induced oxidation without sensitizer produce 9- and 13-LOOH from linoleic acid, which are decomposed to hexanal, 13-oxo-9,11-tridecadienoic acid, 9-oxononanoic acid, caprylic acid, and 2,4-decadienals as major components. This is demonstrated in Figure 2 and Table I. Methyl linoleate was dissolved in ethanol/water and irradiated with light at 25°C in the presence of oxygen during 5 to 20 hr. The volatiles were isolated and investigated by capillary gas chromatography-mass spectrometry, and preparative gas chromatography-infrared spectroscopy. The results and methods used will be published in detail (Bahri and Tressl, 1981). Similar results were recently presented by Chan et al. (8) who injected 13- and 9-LOOH in the injection part of a

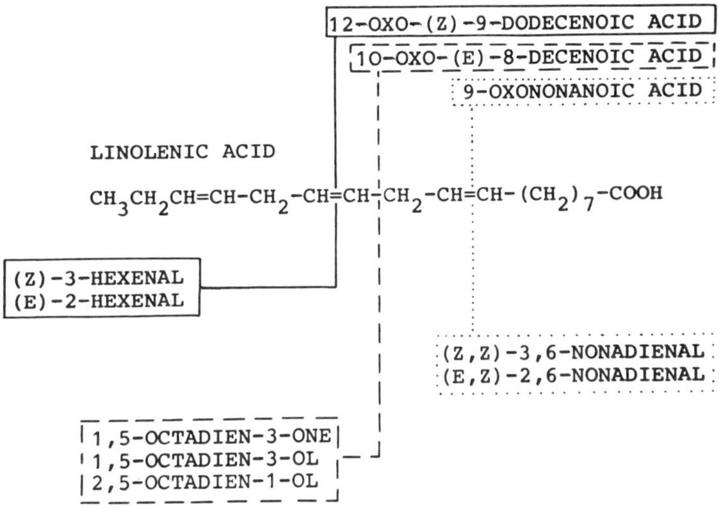

Figure 1. Various components formed from linolenic acid in certain plants, fruits, and vegetables

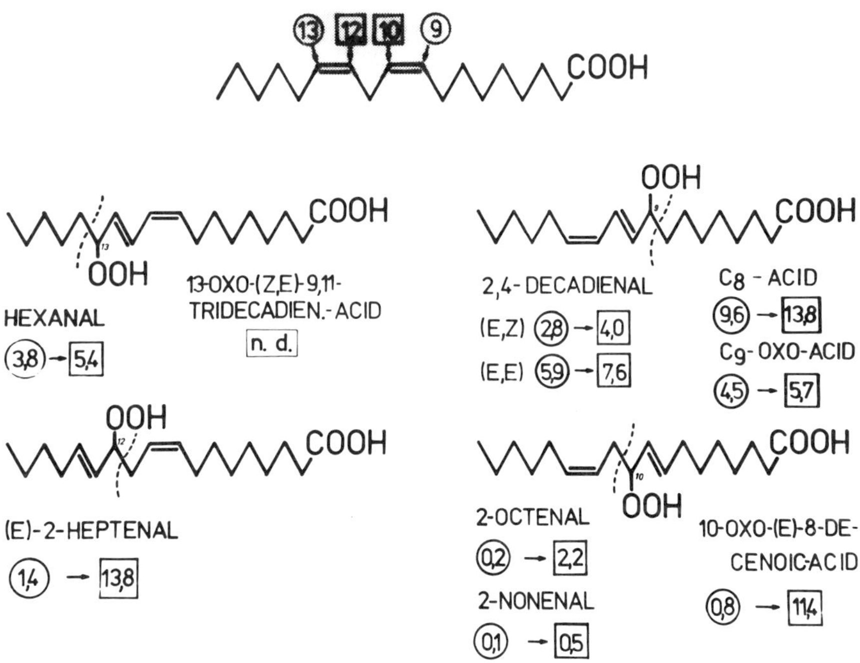

Figure 2. Thermal oxidation, autoxidation, and light-induced oxidation without sensitizer

Table I Formation of Volatiles by Light Induced Oxidation of Methyl Linoleate (ML)

	Component	I_K CW20M (OV 101)	ML+($^3O_2$) µg/mg ML	ML+($^1O_2$) µg/mg ML	Precursor MLOOH
1	Hexanal	1060	3,8	5,4	13-MLOOH
2	2-Pentylfuran	1205	0,73	1,6	10-MLOOH
3	1-Octen-3-on	1286	-	0,4	10-MLOOH
4	(*E*)-2-Heptenal	1296	1,38	13,8	12-MLOOH
5	Methyl caprylate	1372	9,6	13,8	9-MLOOH
6	1-Octen-3-ol	1394	0,4	1,3	10-MLOOH
7	2-Octenal 3-Nonenal	1404 1414	0,2 -	2,3 0,7	10-MLOOH 10-MLOOH
8	2-Nonenal	1510	0,08	0,48	10-MLOOH
9	(*E,Z*)-2,4-Decadienal	1743	2,8	4,0	9-MLOOH
10	(*E,E*)-2,4-Decadienal	1781	5,9	7,6	9-MLOOH
11	Methyl 8-oxooctanoate	1925	0,7	0,95	9-MLOOH
12	Methyl 9-oxononanoate	2010	4,5	5,7	9-MLOOH
13	Methyl 8-(2'-furyl)-octanoate	2097	0,08	0,25	12-MLOOH 13-MLOOH
14	Methyl 10-oxo-(*E*)-8-decenoate	2278	0,78	11,4	10-MLOOH
15	Methyl 11-oxo-9-undecenoate	(1780)	0,8	2,8	10-MLOOH 12-MLOOH
16	Methyl 12-oxo-10-dodecenoate	(1982)	0,2	0,9	12-MLOOH 13-MLOOH
17	Methyl 13-oxo-(9,11)-tridecadienoate	(2175)	1,2	2,6	13-MLOOH

a) Methyl linoleate solution was irradiated with light, presence of oxygen, without sensitizer (autoxidation)

b) Methyl linoleate irradiated under same conditions, with methylene blue as sensitizer (photo-oxidation)

GC-system. The results of the analogous photo-oxidation with methylene blue as sensitizer are presented in Figure 2 and Table I. During this reaction oxygen is transformed into the singlet state, which may attack $>C = C<$ double bonds as an electrophylic agent (9). Therefore, 9-, 10-, 12-, 13-LOOH are formed in equal amounts (10). It can be seen that (E)-2-heptenal amounts from 1.4 to 13.8 mg/g linolenate and 10-oxo-(E)-8-decanoic acid from 0.8 to 11.4 mg/g linoleate. Both constituents may be formed by an allylic fragmentation from 12-LOOH resp. 10-LOOH as shown in Figure 2 (indicating $^1O_2$-oxidation). 2-Octenal and 2-nonenal were determined at 2.3 and 0.48 mg/g linolenate. 1-Octen-3-one and 1-octen-3-ol at 0.4 and 1.3 mg/g. 10-oxo-(E)-8-decenoic acid was identified for the first time as an oxidative degradation product from linoleic acid. It is a homologue of "traumatin" (12-oxo-(E)-10-dodecenoic acid) which was detected as a minor component. Figure 3 presents the mass spectrum of methyl 10-oxo-(E)-8-decenoate acid which has been characterized as an enzymic degradation product from linoleic acid in mushroom homogenates (Tressl et al., 1981). In addition we characterized 8-(2'-furyl)-octanoic acid (which may be derived from 12-oxo-(Z)-9-dodecenoic acid) for the first time as a linoleate product. Figure 4 shows the mass spectrum of methyl 8-(2'-furyl)-octanoate. It was also detected in aroma concentrates of homogenized beans. The results demonstrate the oxidative splitting of linoleate into carbonyls and oxoacids. (E)-2-nonenal and (Z)-3-nonenal, which are known as important flavor components formed by the enzymic degradation were detected as minor components. On the other hand 2-nonenal was formed from oleic acid by photo-oxidation at much higher levels (11). The results show that the thermal fragmentation of 9-LOOH and 13-LOOH, as well as the photo-oxidation of linoleic acid (as discussed by several authors) don't form aldehydes and oxoacids comparable to the enzyme system in plants and fruits.

The results of the analogous experiments with linolenate are summarized in Table II. Major products formed by $^3O_2$-oxidation: caprylic acid, 2,4-heptenals, 2,4,7-decatrienals, 9-oxononanoic acid. Major components indicating $^1O_2$-reactions: 2-butenal, 2,4-heptadienal, and 10-oxo-(E)-8-decenoate. 2-Hexenal, 3-hexenal, 2,6- and 3,6-nonadienals - known as enzymic degradation products from linolenic acids in fruits and vegetables - were detected as minor components.

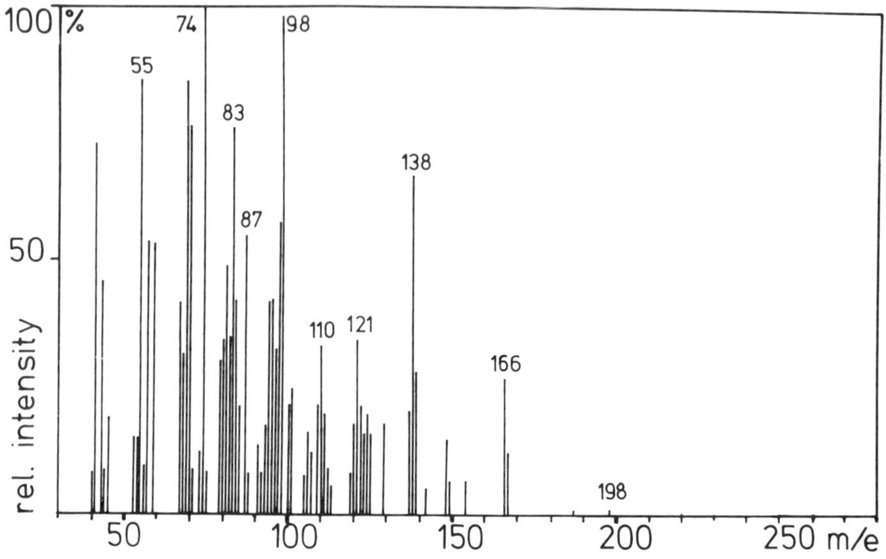

Figure 3. Mass spectrum of methyl 10-oxo-(E)-8-decenoate

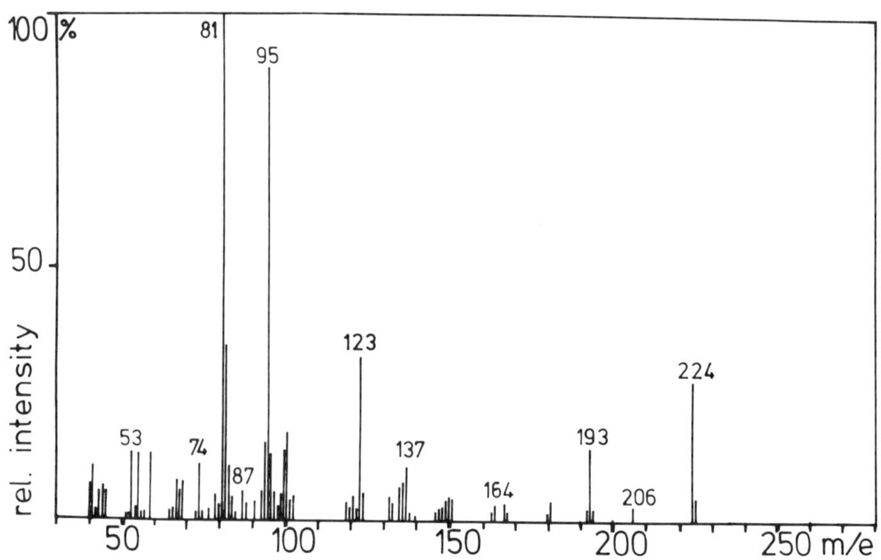

Figure 4. Mass spectrum of methyl 8-(2′-furyl)octanoate

Table II Formation of Volatiles by Light Induced Oxidation of Methyl Linolenate (MLn)

	Component	I_K CW20M (OV 101)	MLn+($^3O_2$) µg/mg MLN	MLn+($^1O_2$) µg/mg MLn	Precursor MLnOOH
1	2-Butenal	1024	0,08	10,7	15-MLnOOH
2	1-Penten-3-ol 3-Hexenal	1122 1136	0,11	0,65 0,05	12-MLnOOH 13-MLnOOH
3	2-Hexenal	1174	0,03	0,20	13-MLnOOH
4	2-Pentenol	1254	0,12	0,60	13-MLnOOH
5	Methyl carpylate	1372	2,0	10,4	9-MLnOOH
6	(E,Z)-2,4-Heptadienal	1435	1,1	7,3	12-MLnOOH
7	(E,E)-2,4-Heptadienal	1467	0,37	2,5	12-MLnOOH
8	3,5-Octadien-2-one	1480	0,04	0,35	9-MLnOOH
9	2,4-Nonadienal	1494	0,03	0,35	9-MLnOOH
10	1,5-Octadien-3-ol	1627	0,05	0,55	10-MLnOOH
11	Methyl undecenoate	1726	0,09	0,86	12-MLnOOH
12	(E,Z,Z)-2,4,7-Decatrienal	1780	0,14	2,0	9-MLnOOH
13	(E,E,Z)-2,4,7-Decatrienal	1845	0,21	2,3	9-MLnOOH
14	(E,E,E)-2,4,7-Decatrienal	1920	0,26	0,43	9-MLnOOH
15	Methyl 9-oxononanoate	2010	5,5	4,3	9-MLnOOH
16	Methyl 8-(2'-furyl)-octanoate	2097	0,17	6,23	10-MLnOOH 13-MLnOOH
17	Methyl 10-oxo-(E)-8-decenoate	2278	0,10	0,45	10-MLnOOH
18	Methyl 11-oxo-9-undecenoate	(1780)	0,05	1,9	10-MLnOOH 12-MLnOOH
19	Methyl 12-oxo-10-dodecenoate	(1982)	-	0,2	12-MLnOOH 13-MLnOOH
20	Methyl 13-oxo-(9,11)-tridecadienoate	(2175)	0,8	1,6	13-MLnOOH
21	Methyl 15-oxo-(9,11,13)-pentadecatrienoate	(2280)	n.d.	1,8	15-MLnOOH

Enzymic Formation of C_6- and C_{12}-Components in Leaves and Fruits

Drawert et al. (1) reported the enzymatic formation of 2-hexenal from linolenic acid in apple, banana, pear, plum, and grape homogenates. Tressl and Drawert (2) demonstrated the enzymic conversion of (^{14}C)-linolenate into (E)-2-hexenal and 12-oxo-(E)-10-dodecenoic acid in homogenates of ripe bananas. In Table III the levels of C_6-components in certain fruits are presented. They are formed during 5 min homogenization at pH 6.8 in the presence of oxygen. It can be seen that apples produce C_6-components depending on the variety. This has been proved with radiolabeling experiments. (E)-2-hexenal is an important flavor constituent in apple aroma concentrates and in apple juice. Some varieties like Glocken produce (Z)-3-hexenal (low activity of (Z)-3 / (E)-2-isomerase E_3). Brambles produced the highest amounts of (E)-2-hexenal but no (Z)-3-hexenal (high activity of E_3). The determined C_6-components are contributing to raspberry flavor and to the aroma of grape juice ("green notes").

Figure 5 shows a reaction scheme which may explain the formation of C_6- and C_{12}-components in leaves and fruits. Hatanaka et al. (12) proposed a cleaving system operative in *Thea sinensis* leaves and other plants which is located in the chloroplasts. E_2 shows a similar substrate specifity as lipoxygenase E_1. The enzymic breakdown products from linolenic acid are (Z)-3-hexenal (I) and 12-oxo-(Z)-9-dodecenoic acid (V). Both constituents are transformed into the corresponding (E)-2-enals by E_3 and/or by chemical reactions. During these transformation the carbonyls may be reduced to alcohols by alcohol oxidoreductase E_4. In ripe fruits the right pathway E_1, E_2, E_3 seems to become operative.

Formation of Carbonyls and Oxoacids in Vegetables and Cereals

This pathway has been demonstrated in many vegetables. In addition to fruits the 9-LOOH are also decomposed into C_9-components by specific endogenous cleaving enzymes E_2 (*Cucurbitaceae*). The situation in vegetables is more complex compared to fruits, since vegetables are normally consumed in cooked form. The breakdown products chemically formed of LOOH are also involved (examples: asparagus (13), potatoes (14), tomatoes (15), beans (16). Table IV presents the formation of C_6-components in raw beans. According to

Table III Formation of C_6-Components in Fruits (ppm)

	Apples		Brambles	Raspberries	Grapes (Riesling)
	Cox	Glocken			
Hexanal	0,5	0,3	0,88	0,26	0,84
Hexanol	0,05	0,02	0,52	0,22	0,10
(Z)-3-Hexenal	0,02	0,50		0,38	0,07
(E)-2-Hexenal	3,9	0,02	7,2	1,0	3,3
(Z)-3-Hexenol	0,01	0,10		0,62	0,13
(E)-2-Hexenol	0,10	0,03	1,7	0,02	0,12

Table IV Formation of C_6-Components in Beans (*Phaseolus vulgaris*) (mg/300 g)

Components	I	II a	II b
Hexanal	0,24	0,33	0,13 0,40
(Z)-3-Hexenal	0,20	0,82	0,46
1-Penten-3-ol	0,05	0,13	0,10
(E)-2-Hexenal	1,95	4,45	3,10
(Z)-3-Pentenol	0,06	0,15	0,21
1-Hexanol	1,30	1,70	1,85
(Z)-3-Hexen-1-ol	0,38	0,50	0,40
(E)-2-Hexen-1-ol	0,40	1,03	0,62
1-Octen-3-ol	0,01	0,02	0,02
12-Oxo-10-dodecenoic acid	0,70	3,80	2,12
8-(2'-furyl)-octanoic acid	+	0,60	0,56

I: Control experiment with fresh beans (300 g)
IIa: Addition of free linolenic acid (100 mg/300 g)
IIb: Addition of methyl linolenate (100 mg/300 g)

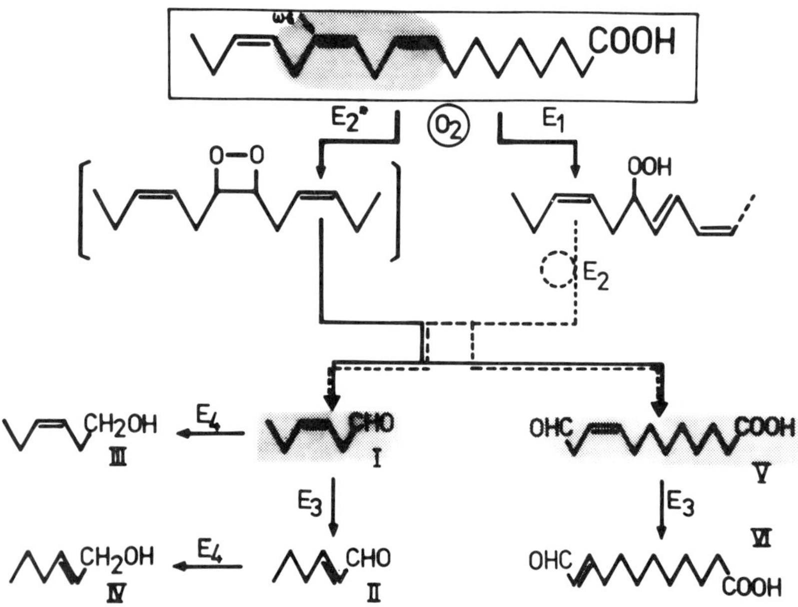

Figure 5. Possible reaction scheme to explain the formation of C_6 and C_{12} components in leaves and fruits

Lumen et al. (17) beans possess a very high lipoxygenase activity. We therefore investigated the formation of C_6-components from linolenic acid in homogenates of beans. The volatiles were determined according to Tressl and Drawert (2). It can be seen that linoleic acid and methyl linolenate are converted into (Z)-3-hexenal, (E)-2-hexenal, and (E)-2-hexenol. Beans possess a very active fatty acid acyl hydrolase activity (E_0). During 1 min homogenization 100 mg methyl linolenate were hydrolized. As shown by Lumen et al. (17) bean seedlings convert linolenic acid into 1-penten-3-ol, (Z)-3-pentenal, (Z)-3-pentenol, and 1-octen-3-ol. Matthew and Galliard (18) determined hexanal and 12-oxo-10-dodecenoic acid as the major C_{12}-component in radiolabeling experiments with (^{14}C)-linoleic acid. In addition we characterized 8-(2-furyl)-octanoic acid for the first time in the aroma extracts of beans.

In Table V some results with cucumbers (*Cucumis sativus*) are summarized. The enzyme system has been studied by Phillips and Galliard (19) and by Phillips et al. (20). Cucumbers possess a lipoxygenase system which forms D-9-LOOH and L-13-LOOH which are decomposed into (E)-2-isomerase (E_3) which converts (Z)-enals into (E)-enals. Therefore, the "labile" (E)-3-enols were detected by Hatanaka for the first time in 1975 (21). They are the precursors of other C_9-constituents in cucumber aroma concentrates.

Experiments I to III were carried out under comparable conditions (250 g cucumbers, phosphatebuffer pH 6.8, 40 s homogenization, addition of 100 mg precursors, volatiles were enriched by extraction / or distillation-extraction and determined by GC-MS). It can be seen that linolenic acid is transformed into hexanal, (Z)-3-nonenal, pentylfuran, and (E)-2-nonenal in the 40 s trial I and I' (I = extraction; I' = distillation-extraction). During 120 s homogenization the labile intermediate (Z)-3-nonenal decreased from 0.23 mg to 0.02 mg and all carbonyls are reduced to some extent to the corresponding alcohols.

Linolenic acid is better metabolized than linoleic acid (6 % : 15 %). (Z,Z)-3,6-Nonadienal is transformed into (E,Z)-2,6-nonadienal and pentenylfuran, which has been characterized for the first time. In all experiments 9-oxononanoic acid was formed as a major component as shown recently (22). Green bananas possess a similar enzyme system producing strictly C_9-components.

Figure 6 shows the transformation of (Z)-3-nonenal into aroma constituents which were characterized in

Table V Formation of C_9-Components in Cucumber (*Cucumis sativus*)

		Precursor				
		Linoleic acid			Linolenic acid	
		I	I'	II	III	III'
1	Hexanal	0,35	0,75	0,08	-	-
2	(*Z*)-3-Nonenal	0,23	0,20	0,02	0,03	0,06
3	Pentylfuran	0,35	0,32	0,11	0,06	0,10
4	(*E*)-2-Nonenal	3,0	3,12	0,72	0,40	0,60
5	(*E*)-2-Hexenal	-	-	-	0,03	0,11
6	(*Z,Z*)-3,6-Nonadienal	0,07	-	-	0,15	0,11
7	Pentenylfuran	-	-	-	0,04	0,06
8	Unidentified carbonyl	-	-	-	0,15	0,60
9	(*E,Z*)-2,6-Nonadienal	1,1	1,05	0,28	6,2	6,9

I 40 s homogenization; extraction 24 h
I' 40 s homogenization; extraction-distillation 2 h
II 120 s homogenization; extraction-distillation 2 h
III 40 s homogenization; extraction 24 h
III' 40 s homogenization; extraction-distillation 2 h

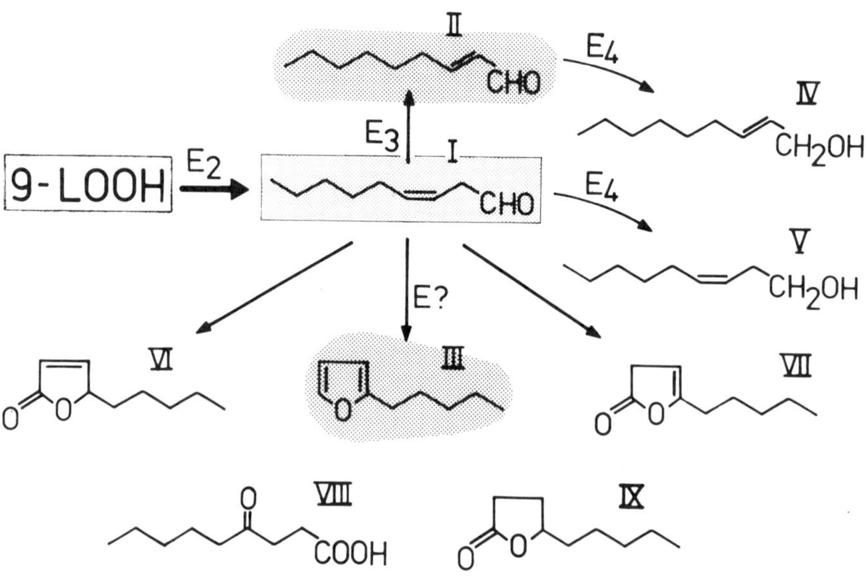

Figure 6. Transformation of (Z)-3-nonenal into aroma constituents

cucumber homogenates and in model systems. (Z)-3-Nonenal is converted into (E)-2-nonenal, 2-pentylfuran and the corresponding alcohols by the enzymes E_3 and E_4. In acidic media (pH 3 - 4) this reactive precursor is transformed into the (E)-enal, pentylfuran and the lactones VI, VII and IX. In an analogous reaction (Z,Z)-3,6-nonadienal is converted into the corresponding C_9-constituents.

The enzymic formation of C_6- and C_9-components in vegetables is determined by the cleaving-system E_2. Tomatoes possess a lipoxygenase-system which forms 9-LOOH and 13-LOOH in a ratio of 95 : 5 (23). According to Matthew et al. (24) only the 13-LOOH is decomposed into C_6- and C_{12}-components by the lyase-system E_2. This is consistant with the results of Kazeniac et Hall (15) who demonstrated the formation of (E)-2-hexenal, (Z)-3-hexenal and (Z)-3-hexen-1-ol from linolenic acid in tomato homogenates.

In cereal products like wheat and barley malt (E)-2-nonenal was detected as a major component (11). On the other hand (E)-2-nonenal is formed in the trace range by enzymatic reactions in barley and green malt. (E)-2-Nonenal is known as the most important off-flavor constituents in oxidized beer. Therefore, we looked for possible precursors of 2-nonenal. As shown in Figure 7 we prepared 9-hydroperoxy-(E,Z)-10,12-octadienoic acid by incubation of tomato homogenates with linoleic acid. The 9-LOOH was transformed into the corresponding α-ketol with a crude enzyme extract from green malt. Barley, green malt and other cereals contain hydroperoxide isomerase but no cleaving enzymes (25). The α-ketol was reduced to 9,10-dihydroxy-(E)-12-octadecenoic acid. The components presented in Figure 7 were separated by HPLC and characterized by mass spectrometry, infrared- and NMR-spectroscopy. The results will be published in detail (Tressl and Bahri, 1981).

The purified dihydroxy acid was a labile component and decomposed during heating into carbonyls and oxoacids. Figure 8 presents some results of the thermal fragmentation at pH 4 - 5 in a Likens-Nickerson distillation. It can be seen that we characterized constituents similar to those in the linoleate experiment of cucumber homogenates. 50 mg precursor (9,10-dihydroxy-(Z)-1,2-octadecenoic acid) were decomposed into: 1.95 mg (Z)-3-nonenal, 0.25 mg (E)-2-nonenal, 0.51 mg 2-pentylfuran, and C_9-oxo- and C_9-dicarboxylic acids. In our opinion this is the most important reaction in oxidized beer (producing cardboard flavor). In an analogous reaction linolenic acid was trans-

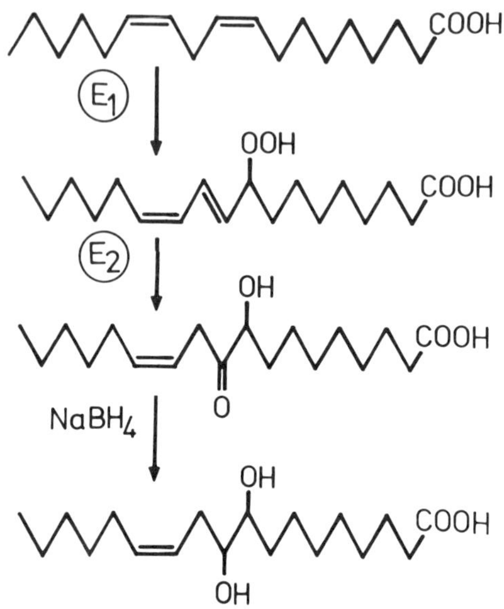

Figure 7. Synthesis of 9-hydroperoxy-(E,Z)-10,12-octadienoic acid

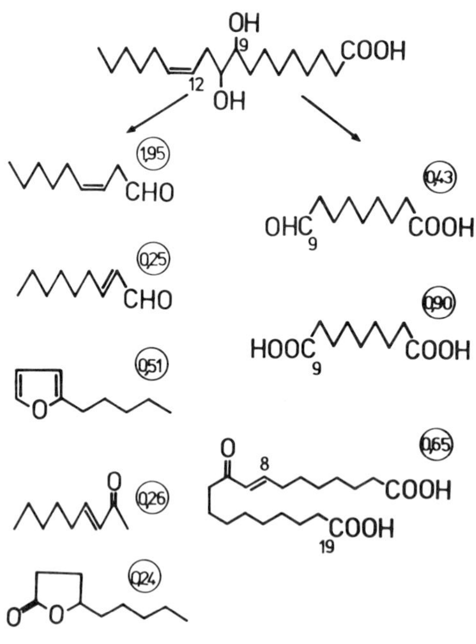

Figure 8. Results of the thermal fragmentation at pH 4–5 in a Likens–Nickerson distillation

formed into the corresponding dihydroxy acid. During this reaction we observed the formation of phenols and aromatic hydrocarbons and some not identified products. But 3,6-nonadienal and 2,6-nonadienal were formed as major components. Vick and Zimmerman (26) demonstrated a fatty acid cyclase in cereals catalizing the conversion of 13-LnOOH into cyclopentenoic acid, which have similar structures as prostaglandines.

Formation of C_8- and C_{10}-(C_{11})-Components in Edible Mushrooms (Agaricus campestris)

Edible mushrooms like *Agaricus bisporus* (27). *Lentinus edodes* Sing (28), *Boletus edulis* (29) produce 1-octen-3-ol, 3-octanol, 2-octen-1-ol, and 1-octen-3-one as volatile constituents. 1-Octen-3-ol possesses a mushroom-like aroma and is known as "mushroom alcohol". Freytag and Ney (30) isolated this component from aroma concentrates and demonstrated a laevorotatory form in mushrooms. The racemic form is known as a product from oxidized linoleate in cereals and vegetables. The optical activity indicates a biosynthetic formation of (-)-1-octen-3-ol in mushrooms. Lumen et al. (17) demonstrated a lipoxygenase-system catalizing the conversion of linoleic acid into 1-octen-3-ol. We investigated the enzymic conversion of linoleic-, and linolenic acids into C_8- and C_{10}-components by mushrooms. The methods used will be published in detail (Tressl et al., 1981). Some of the results are summarized in Table VI. Fresh mushrooms (*Agaricus campestris*) were homogenized with phosphatebuffer at pH 6.8 for 5 min and the volatiles concentrated by distillation-extraction. The volatiles were identified and quantified by capillary GC-mass spectrometry. Major components: 1-Octen-3-ol, (*Z*)-2-octen-1-ol, 3-octanone, and 1-octen-3-one. In an analogous experiment with linoleic acid 1-octen-3-ol, 1-octen-3-one, (*Z*)-2-octen-1-ol, and (*Z*)-2-octenal increased considerably. In addition we characterized C_{10}-components for the first time in mushroom homogenates:

10-Oxodecanoic acid	3,4 ppm
10-hydroxydecanoic acid	32,6 ppm
10-Oxo-8-decenoic acid	1,93 ppm
10-Hydroxy-8-decenoic acid	0,76 ppm
9-Oxodecanoic acid	0,3 ppm
9-Hydroxydecanoic acid	0,5 ppm
and two C_{11}-components	
10-Oxoundecanoic acid	1,38 ppm
10-Hydroxyundecanoic acid	0,76 ppm

Table VI Formation of Volatiles in Mushrooms (*Agaricus campestris*)

	Component	Concentration (ppm) I	II Linoleic acid	III Linolenic acid	Identification
1	Hexanal	0,23	0,56	0,09	R_t, MS
2	1-Octanol	+	+	+	R_t, MS
3	3-Octanone	8,65	7,8	5,5	R_t, MS
4	3-Octanol	1,5	2,0	1,25	R_t, MS
5	1-Octen-3-one	0,7	2,2	1,1	R_t, MS
6	1-Octen-3-ol	30,1	61,0	25,1	R_t, MS
7	(Z)-2-Octenal	1,8	7,5	1,9	R_t, MS
8	(Z)-2-Octen-1-ol	4,8	17,4	5,0	R_t, MS
9	(Z)-1,5-Octadien-3-one	+	+	0,33	R_t, MS, IR
10	(Z)-1,5-Octadien-3-ol	+	+	27,5	R_t, MS, IR
11	(Z,Z)-2,5-Octadienal	-	-	+	R_t, MS
12	(Z,Z)-2,5-Octadien-1-ol	+	+	17,9	R_t, MS, IR
13	Benzaldehyde	15,8	11,7	10,8	R_t, MS

I Control experiment with fresh mushrooms

II Addition of linoleic acid before homogenization

III Addition of linolenic acid before homogenization

All oxo- and hydroxy acids were characterized for the first time as enzymic degradation products from linoleic acid. The major components 1-octen-3-ol and 1,5-octadien-3-ol were optically active. This was proved by formation of an ester with an optically active acid and capillary separation. The methods used will be published in the near future (Tressl and Engel, 1981). These results indicate a highly specific enzyme-system in Agaricus campestris catalyzing the conversion of linoleic acid into (-)-1-octen-3-ol and 10-hydroxydecanoic acid resp. into (Z)-2-octen-1-ol and 9-hydroxydecanoic acid. As demonstrated in Table VI linolenic acid is transformed into the corresponding C_8-components containing two $>C = C<$ bonds. (-)(Z)-1,5-Octadien-3-ol possesses a strong mushroom aroma with an earthy note. (Z,Z)-2,5-octadien-1-ol is perceived mushroom-like and (Z)-1,5-octadien-3-one has a metallic, fungal odour. They were characterized for the first time in mushrooms. Figure 9 shows a possible reaction scheme which may explain the formation of C_8- and C_{10}-components in Agaricus campestris. Mushrooms possess a lipoxygenase-system E_1 and a very active cleaving system E_2. We assume two pathways. 13-LOOH may be converted (via a γ-ketol or γ-ketoepoxide) into 1-octen-3-one, 10-oxodecanoic acid and 10-oxodecenoic acid. 1-Octen-3-one is reduced by an alcohol oxidoreductase (E_4) to the optically active "mushroom alcohol" (-)-1-octen-3-ol and the oxoacids to the corresponding hydroxyacids. The 9-LOOH may be transformed (via a α-ketol or α-ketoepoxide) into (Z)-2-octenal and 9-oxodecanoic acid which are reduced by E_4 to the corresponding alcohols. The formation of 10-oxoundecanoic acid and 10-hydroxyundecanoic acid cannot be explained by this "working hypothesis". Experiments which may prove this are in progress.

Figure 10 summarizes some results of the lipid-oxidation system in basidiomycetes, leaves, fruits, vegetables, and cereals. It can be seen that there is a development with evolution and differentiation. The enzyme system is highly substrate specific to a (Z,Z)-1,4-pentadiene system converting linoleic- and linolenic acids into carbonyls and oxoacids which may undergo further isomerization (E_3) and/or reduction (E_4). Some of the components formed are very potent aromatics, pheromones, and wound hormones. Basidiomycetes and Fungi produce (-)-1-octen-3-ol as their sensorial principle. 9-Oxodecanoic acid is known as queen substance and is the sex pheromone of honey bees.

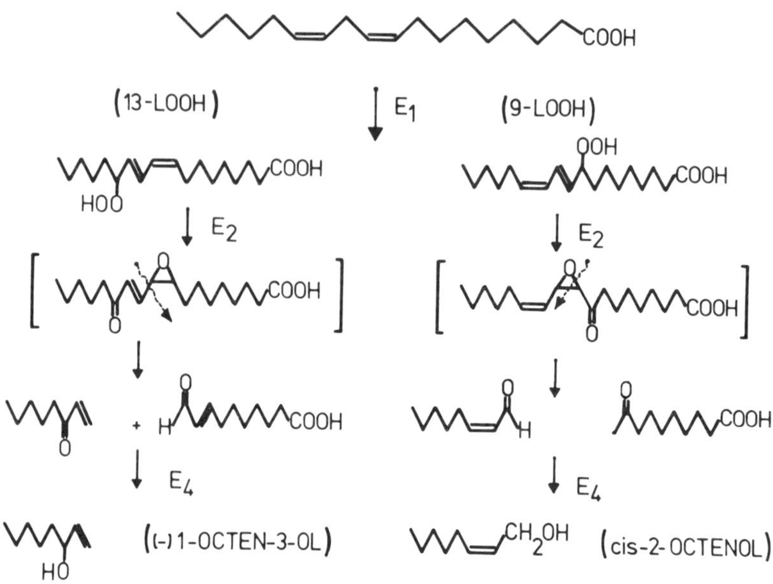

Figure 9. Possible reaction scheme to explain formation of C_6 and C_{10} components in Agaricus campestris

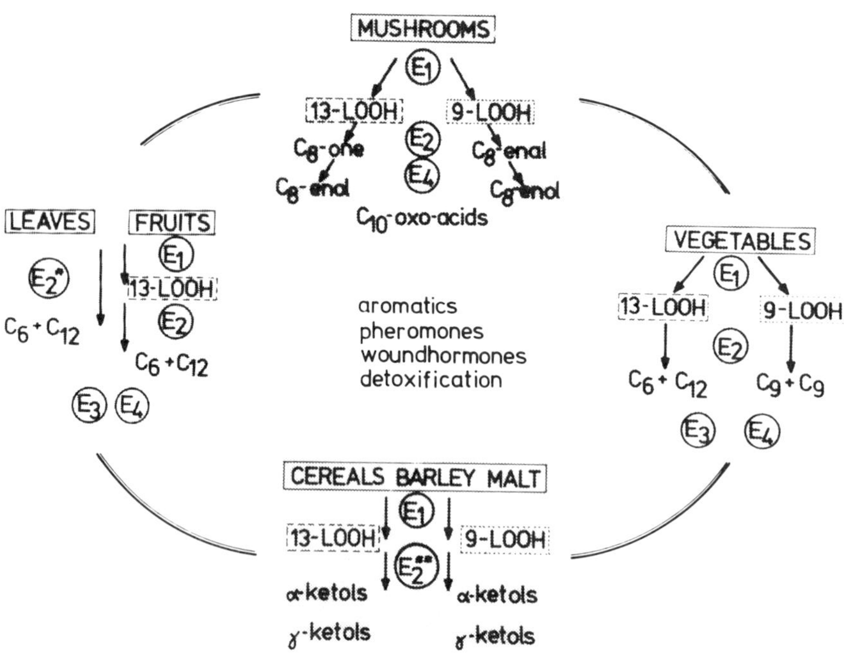

Figure 10. Lipid-oxidation system in basidiomycetes, leaves, fruits, vegetables, and cereals

Leaves and fruits form C_6-components with "green notes" and 12-oxo-Z-9-dodecenoic resp. 12-oxo-E-10-dodecenoic acids which were characterized as the wound hormone traumatin.

In cereals the cleaving enzyme E_2 is transformed into a hydroperoxid-isomerase and into a hydroperoxid cyclase producing prostaglandine-like components.

Literature Cited

1. Drawert, F.; Heimann, W.; Emberger, R.; Tressl,R. Liebigs Ann. Chem. 1966, 694, 200
2. Tressl, R.; Drawert, F. J. Agr. Food Chem. 1973, 21, 560
3. Vick, B.A.; Zimmerman, D.C. Plant Physiol. 1976, 57, 780
4. Galliard, T.; Matthew, J.A. Phytochemistry 1977, 16, 339
5. Wardale, D.A.; Lambert, E.A.; Galliard, T. Phytochemistry 1978, 17, 205
6. Phillips, D.R.; Galliard, T. Phytochemistry 1978, 17, 355
7. Hatanaka, A.; Kajiwara, T.; Sekiga, J.; Kido, J. Phytochemistry 1977, 16, 1828
8. Chan, H.W.-S.; Prescott, F.A.A.; Swoboda, P.A.T. J. AOCS 1976, 53, 1972
9. Schenck, G.O. Angewandte Chemie 1957, 18/19, 579
10. Terao, J.; Ogawa, T.; Matsushita, S. Agr. Biol. Chem. 1975, 39, 397
11. Tressl, R.; Bahri, D.; Silwar, R. "Bildung von Aldehyden durch Lipidoxidation und deren Bedeutung als "off-flavor"-Komponenten in Bier"; Proceedings of the European Brewery Convention Congress, Berlin (West), 1976; p 27
12. Hatanaka, A.; Sekiga, J.; Kajiwara, T.; Miura, T. Agr. Biol. Chem. 1979, 43, 735
13. Tressl, R.; Bahri, D.; Holzer, M.; Kossa, T. J. Agric. Food Chem. 1977, 25, 459
14. Buttery, R.G.; Ling, L.C. J. Agr. Food Chem. 1972, 20, 698
15. Kazeniac, S.J.; Hall, R.M. J. Food Scie. 1970, 35, 519
16. Buttery, R.G. J. Agr. Food Chem. 1975, 23, 1003
17. Lumen, B.O.; Stone, E.J.; Kazeniac, S.J.; Forsythe, R.H. J. Food Scie. 1978, 43, 698
18. Matthew, J.A.; Galliard, T. Phytochemistry 1978, 17, 1043
19. Phillips, D.R.; Galliard, T. Phytochemistry 1978, 17, 355
20. Phillips, D.R.; Matthew, J.A.; Reynolds, J.; Fenwick, G.R. Phytochemistry 1979, 18, 401

21. Hatanaka, A.; Kajiwara, T.; Harada, T. Phytochemistry 1975, 14, 2589
22. Tressl, R.; Holzer, M.; Apetz, M. "Biogenesis of Volatiles in Fruit and Vegetables"; Proceedings Intern. Symp. Aroma Research, Zeist 1975, Pudoc Wageningen, p 41
23. Galliard, T.; Matthew, J.A. Phytochemistry 1977, 16, 339
24. Matthew, J.A.; Chan, H.W.-S.; Galliard, T. Lipids 1977, 12, 324
25. Yabuuchi, S.; Amaha, M. Phytochemistry 1976, 15, 387
26. Vick, B.A.; Zimmerman,D.C. Plant Physiology 1979, 64, 203
27. Picardi, S.M.; Issenberg, P. J. Agr. Food Chem. 1973, 21, 959
28. Kameoka, H.; Higuchi, M. Nippon Nogeikagachu Kaishi 1976, 50, 185
29. Pyysalo, H. Acta Chemica Scandinavica 1976, B 30, 235
30. Freytag, W.; Ney, K.H. J. Biochem. 1968, 4, 135

Acknowledgements

This work was supported by Arbeitsgemeinschaft industrieller Forschungsvereinigungen e.V., Köln.

RECEIVED April 3, 1981.

INDEX

INDEX

O

P

Q

R

Jacket design by Carol Conway.

Elements typeset by Service Composition Co., Baltimore, MD.
Printed and bound by Maple Press Co., York, PA.